*Proceedings of the Thirty-second
Annual Biology Colloquium*

The Annual Biology Colloquium

YEAR, THEME, AND LEADER

1939. *Recent Advances in Biological Science.* Charles Atwood Kofoid

1940. *Ecology.* Homer LeRoy Shantz

1941. *Growth and Metabolism.* Cornelius Bernardus van Niel

1942. *The Biologist in a World at War.* William Broadbeck Herns

1943. *Contributions of Biological Science to Victory.* August Leroy Strand

1944. *Genetics and the Integration of Biological Sciences.* George Wells Beadle

1945. (Colloquium cancelled)

1946. *Aquatic Biology.* Robert C. Miller

1947. *Biogeography.* Ernst Antevs

1948. *Nutrition.* Robert R. Williams

1949. *Radioisotopes in Biology.* Eugene M. K. Geiling

1950. *Viruses.* W. M. Stanley

1951. *Effects of Atomic Radiations.* Curt Stern

1952. *Conservation.* Stanley A. Gain

1953. *Antibiotics.* Wayne W. Umbreit

1954. *Cellular Biology.* Daniel Mazia

1955. *Biological Systematics.* Ernst Mayr

1956. *Proteins.* Henry Borsook

1957. *Arctic Biology.* Ira Loren Wiggins

1958. *Photobiology.* F. W. Went

1959. *Marine Biology.* Dixy Lee Ray

1960. *Microbial Genetics.* Aaron Novick

1961. *Physiology of Reproduction.* Frederick L. Hisaw

1962. *Insect Physiology.* Dietrich Bodenstein

1963. *Space Biology.* Allan H. Brown

1964. *Microbiology and Soil Fertility.* O. N. Allen

1965. *Host-Parasite Relationships.* Justus F. Mueller

1966. *Animal Orientation and Navigation.* Arthur Hasler

1967. *Biometerology.* David M. Gates

1968. *Biochemical Coevolution.* Paul R. Erlich

1969. *Biological Ultrastructure: The Origin of Cell Organelles.* John H. Luft

1970. *Ecosystem Structure and Function.* Eugene P. Odum

1971. *The Biology of Behavior.* Bernard W. Agranoff

1972. *The Biology of the Oceanic Pacific.* John A. McGowan

The Biology of Behavior

The Biology of Behavior

Proceedings of the Thirty-second
Annual Biology Colloquium

Edited by JOHN A. KIGER, JR.

Corvallis:

OREGON STATE UNIVERSITY PRESS

Library of Congress Cataloging in Publication Data

Biology, Colloquium, 32nd, Oregon State University, 1971.
  The biology of behavior.

  (Proceedings of the Annual Biology Colloquium, 32nd)
  Includes bibliographies.
  1. Psychobiology—Congresses.   2. Animals, Habits
and behavior of—Congresses.   3. Human behavior—
Congresses.   4. Neurochemistry—Congresses.   I. Kiger,
John A., ed.   II. Title.   III. Series: Biology
Colloquium, Oregon State College. Annual Biology
Colloquium proceedings, 32nd.
QH301.B43   32nd   [QL785]   574'.08s   [591.5]
ISBN 0-87071-171-7   72-7409

Printed in the United States of America

Contents

Preface

IN RECENT YEARS it has become apparent that the study of behavior is about to undergo a revolution, if indeed the revolution has not already begun. Indicative of this view, is the air of confidence exuded by workers in the field that questions which were once considered unanswerable are becoming tractable and that questions which were once experimentally inconceivable are being asked. Indicative also of the revolution is the influx of new minds to the field-scientists trained in the disciplines of biochemistry, genetics, electrical engineering, and computer sciences—augmenting the more traditional disciplines of psychology and psycobiology. A confidence is growing that even if the human mind can never fully understand its own depths those depths are measurable, and definable in terms of physical and chemical parameters.*

To a large extent, this new confidence is an outgrowth of the phenomenal successes experienced in the field of biology in the last twenty years. The Watson-Crick model for the structure of DNA followed by the elaboration of the genetic code has provided a framework within which we can understand our existence as self-perpetuating aggregations of molecules. The really major question that thus far eludes us is how an aggregation of molecules, albeit an extremely complex one, such as the human brain can exhibit that state which we call consciousness. It is an article of faith, that the answer to this question can be comprehended by the object of the question, yet the search for the answer as well as the elucidation of peripheral issues is one of the most exciting adventures in science today.

The purpose of this 32nd Annual Biology Colloquium was to assess the current state of behavioral studies. The accuracy of such an assessment is limited of course by the restrictions of both time and choice. Participants were invited in the belief that their areas of research would be particularly significant in future advances in the field.

---

* Sinsheimer Robert L., 1971, "The Brain of Pooh: An Essay on the Limits of Mind," *American Scientist 59*, 20-28.

Only history will be able to confirm this belief. At least, it is hoped, these Proceedings will stand as a record of work deemed exciting and significant at this point in our progress.

As chairman of the Colloquium Committee I would like to take this opportunity to thank the other members of the Colloquium Committee as well as Dr. Bernard Agranoff, Colloquium Leader, and the speakers for their efforts at making this a successful colloquium.

JOHN A. KIGER, JR.
Oregon State University

Biochemical Concomitants
of Learning and Memory

BERNARD W. AGRANOFF
Department of Biological Chemistry and
Mental Health Research Institute
University of Michigan
Ann Arbor, Michigan 48104

THIS PRESENTATION will deal both with some generalizations concerning biochemical approaches to behavior and to actual experimental results. Figure 1 is a diagrammatic representation of a neuron as seen by a biochemist. As with other eucaryotic cells, DNA and RNA synthesis are concentrated within the nucleus, while protein synthesis takes place primarily in the cytoplasm on ribosomal aggregates. In the case of the neuron, the ribosomes are believed to be exclusively in the perikaryon surrounding the nucleus. Neurons differ from other cells in that they may be extremely long, and since it is our present understanding that neuronal protein synthesis takes place only near the nucleus, proteins which are to be effective at some distance, such as at the synapse, must be transported to the site of action. This refers to the process of axonal (or axoplasmic) flow.

Figure 1 also indicates the site of action of several antibiotics. These substances, which have arisen by natural selection and mutation, are nature's metabolic monkey wrenches which disrupt growth processes in the microenvironment of organisms that secrete them. They are also effective as blocking agents in higher animals, and they are particularly attractive because of their specificity of action. For example, DNA synthesis can be blocked by arabinosyl cytosine, RNA synthesis by actinomycin D, and protein synthesis by the cycloheximide antibiotics or by puromycin. It then becomes possible to measure differential effects on behavior by a block of each of these macromolecular processes.

It has been learned over the past 50 years that metabolic blocking agents which disrupt energy conversion processes in the brain are not useful in behavioral studies because they produce gross neurological abnormalities, such as convulsions and unconsciousness, and many of them irreversibly damage the nervous system. For example, it is well

1

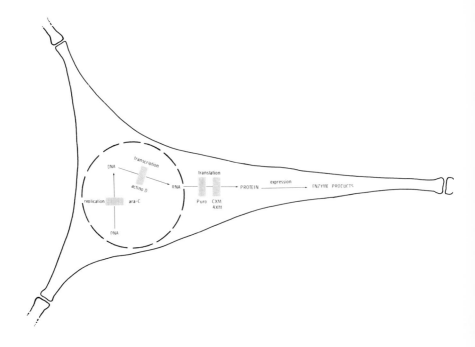

FIGURE 1. Diagrammatic representation of a vertebrate neuron. DNA and RNA synthesis are localized in the nucleus. Replication of neurons is not believed to occur in adult animals. DNA synthesis is blocked by arabinosyl cytosine (ara-C), while DNA-dependent RNA synthesis is blocked by actinomycin D (actino D). RNA-directed protein synthesis occurs in the ribosomes, and it is localized to the region surrounding the nucleus. Newly synthesized proteins migrate in a somatofugal fashion to the extremities of the cell, often at a great distance *via* the process of axoplasmic flow. Protein synthesis can be blocked *in vivo* by glutarimide antibiotics, such as cycloheximide (CXM) or acetoxycycloheximide (AXM).

known that disruption of oxygen or glucose supply to the brain can rapidly lead to death. Since the macromolecular synthesis-blockers do not grossly disrupt metabolism, the question arose as to what behavioral effects the latter agents might produce. Studies by Flexner (1967) in the mouse indicated that intracerebral puromycin injections blocked memory of a recently learned task. This finding served as a starting point for our research. We have concerned ourselves primarily with protein inhibitors such as acetoxycycloheximide and puromycin (Figure 2). Acetoxycycloheximide is an extremely potent inhibitor of protein synthesis and blocks peptide formation. Puromycin attaches to the

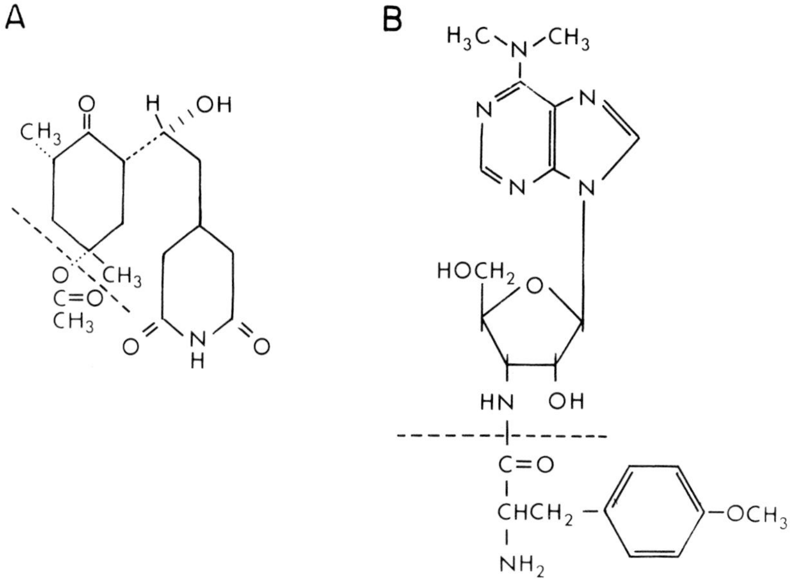

A

B

FIGURE 2. Structures of protein blockers. A. Acetoxycycloheximide, an extremely potent inhibitor of protein synthesis. Replacement of the acetoxy group (below the dotted line) by H gives the structure of a second, less potent antibiotic, cycloheximide. B. Puromycin. Hydrolysis across the amide bond shown at the dotted line yields methyl tyrosine and puromycin aminonucleoside, an analog of puromycin which is inactive in the block of protein synthesis. The complete molecule blocks protein synthesis by mimicking the amino-acyl terminus of transfer RNA.

growing peptide chain which is then released from the ribosome. While both agents block protein synthesis, acetoxycycloheximide slows down the peptide synthesizing process, while puromycin chops the peptides off before the proteins are completed. The resemblance of puromycin to amino-acyl RNA is well known (Yarmolinsky and De la Haba, 1959), and the dotted line in Figure 2 separates the nucleotide moiety from the amino acid analog.

The Goldfish as an Experimental Animal

We have used the antibiotics to selectively block brain protein, RNA, and DNA synthesis in the goldfish. This animal is especially convenient for various reasons. The brain is enclosed by rather primitive membranes in a relatively roomy cavity so that it is accessible to permeation by injected drugs. A 10-gram goldfish has a brain weighing about 80 milligrams. The optic tracts are completely crossed and project upon the contralateral optic tectum. The forebrain is relatively undeveloped. To inject radioactive isotopes or the antibiotics, we have developed an injection technique using a Hamilton syringe and a 30-gauge needle (Agranoff and Klinger, 1964). The unanesthetized animal can readily be injected with 10 microliters of a solution which is then distributed within the cranial cavity. In experiments reported elsewhere we have found that the injection of 170 micrograms of puromycin, 10 micrograms of cycloheximide, or 0.2 micrograms of cycloheximide (Agranoff, 1967) produces a rapid inhibition of brain protein synthesis to less than 20 percent of normal, with recovery within a day. Actinomycin D selectively blocks RNA synthesis, and although it should have no effect on protein synthesis, we find in fact that there is a secondary inhibition of protein synthesis beginning several hours after the inhibition of RNA synthesis (Agranoff et al., 1967). Behavioral effects have not yet been discussed; however, in the course of these biochemical experiments, we had already made a striking observation: while a severe block in DNA, RNA, or protein synthesis was in effect, there were no detectable behavioral abnormalities—fish swam about normally.

Shock-Avoidance Training

Behavioral training

We have trained goldfish to swim over a barrier following the onset of a light signal in order to avoid a mild punishing shock applied through the water. The training apparatus consists of a long tank with a barrier in the center, dividing it into two similar compartments. At the beginning of a trial, a light goes on on the side of the apparatus in which the fish has been placed, and 20 seconds later intermittent punishing shock is administered through the water. Without prior training, the fish swims over to the dark, safe-appearing side of the apparatus. The next trial begins with onset of light on the side of the apparatus to which the fish has escaped. Depending on the details of construction of the apparatus and the time interval between the onset of the light stimulus and the unconditioned stimulus (the punishing shock), we see varying rates of acquisition of an avoidance response— swimming over the barrier before the onset of shock. Generally, sufficient training can be given such that following 20 1-minute trials on the first day of an experiment, fish will demonstrate memory (show an avoidance-responding rate significantly higher than that of untrained fish) days, weeks, or months later when retested in the apparatus. Our standard paradigm consisted of giving fish 20 trials on Day 1 of an experiment and 10 retraining trials three days later. We found considerable variability between fish and even for groups of fish trained at different times of the year, but by means of a regression equation which predicted Day 4 scores on the basis of Day 1 scores, we were able to normalize and compare scores.

Studies with antibiotics

We first learned that fish given 20 training trials and then injected with puromycin showed no memory of what they had learned when tested three days later (Agranoff, 1967). Fish given 20 trials and returned to home storage tanks, then given puromycin an hour later, showed the responding rate expected for control animals. It therefore became evident that puromycin produced an effect on performance but only when given immediately following training. It appeared that memory of training was "solidified" or *consolidated* following training and that the part of the process that was susceptible to puromycin was completed within an hour. Many ancillary confirming sorts of experiments were performed (Agranoff, 1967). A particularly significant one involved the injection of puromycin before training. As noted above, there is little gross disruption in behavior following injection of antibiotic. We noted that fish injected with the antibiotic intracranially learned at a normal rate, but upon retesting three days later, they showed a profound memory deficit. These results suggested to us that memory was formed by a multiphasic process. The acquisition phase, or learning, was unaffected by the antibiotic, while the storage or conversion into a long-term form was blocked if the antibiotic were injected during some critical period. We also learned that following injection of the antibiotic, memory was not lost immediately, but appeared to decrease over a period of two to three days. We therefore must distinguish whether the antibiotic causes the formation of a faulty memory or gives rise to some sort of interference process which takes time to "develop." We also learned that if fish are allowed to remain in the training apparatus following training, the susceptibility to puromycin is extended. The consolidation process appears to begin when the animals have been returned to their home tanks. The possible implications of these results are described elsewhere (Agranoff and Davis, 1968; Davis and Agranoff, 1966).

Studies with other antibiotics

Essentially identical results were obtained when acetoxycycloheximide or actinomycin D, inhibitors of protein synthesis and RNA synthesis, respectively, were used instead of puromycin. When puromycin aminonucleoside, a derivative of puromycin that does not block protein synthesis, was used, there was no effect on memory (Agranoff, 1966). Arabinosyl cytosine, an inhibitor of DNA synthesis, produced no effect on either acquisition or on memory (Casola et al., 1968).

Mechanism of antibiotic action on behavior

Originally, the antibiotics were used in these experiments in the hope that a macromolecular role in memory could be pinpointed. From the temporal standpoint, these experiments have been successful. Anatomical localization of the effects within the brain has not yet proven possible, although there have been suggestions that in the mouse, for the first few days following training, puromycin is effective in the temporal cortex but then diffuses through the entire brain (Flexner, Flexner, and Roberts, 1967). A question that has been discussed widely is whether the antibiotics are producing their effects *via* their known biochemical roles. For example, does puromycin block memory by blocking protein synthesis, or by some side effect? Our studies with puromycin aminonucleoside would seem to suggest that it is the protein-inhibiting effect of puromycin that blocks memory, since other known effects of puromycin such as on cyclic AMP (Appelman and Kemp, 1966) or its convulsion-potentiating effects are shared with the memory-inactive substance, puromycin aminonucleoside (Agranoff, 1970). Also, the various antibiotics bear little chemical resemblance to each other and have in common only macromolecular blocking effects.

Another sort of issue however is the following: Given that puromycin blocks memory *via* inhibition of protein synthesis, is there a specific protein synthetic step being blocked or is there some nonspecific toxic process in the brain, resulting from decreased protein synthesis, that selectively blocks memory? Such questions are difficult to answer, but we have made attempts in this direction. For example, we now know that if fish are anesthetized with a fish anesthetic immediately following training, and are then injected with a protein blocker, possible unpleasant sensations resulting from the injection might not be experienced. Under such circumstances, we find that a memory block is still obtained (Agranoff, 1971). When, however, fish are cooled, a resultant inhibition of protein synthesis is seen by decreased incorporation of amino acid precursors into protein, but there is no memory loss (Neale and Gray, 1971). In order to produce the disruptive effect on memory, normal metabolism must be going on while macromolecular processes are being blocked.

Other Biochemical Studies on Learning and Memory

The use of antibiotics to study behavior is only one example of the interaction of chemical technology with behavior. Other laboratories have been examining the effect of training on incorporation of various RNA and protein precursors. In the mouse, it has been reported that training causes an increased labeling of RNA from uridine, a precursor of RNA *via* a salvage pathway (Glassman, 1969). Similar increases in protein synthesis have, in general, not been reported. In an attempt to look for a slowly turning-over protein, Lajtha and Toth (1966) injected radioactive precursors into pregnant rats. Within a few days following birth, very little radioactive protein remained in the brains of the newborn, suggesting that virtually all of the brains' proteins turn over. Other laboratories, such as that of Hydén, have attempted to localize changes in protein and small brain regions believed to be associated with a specific behavior. Alterations in both labeling protein and in the distribution of protein in the hippocampus of rats trained in a handedness task have been reported (Hydén and Lange, 1970).

Subsequent papers will report on a wide sampling not only of other biochemical techniques but behavioral, genetic, and electrophysiological attempts to understand what brain and behavior have in common. From the outset, we acknowledge the difficulties that the experimentalist in this area of neuroscience encounters. However, we hope also to impart some of the enthusiasm we have for the questions we ask and the faith we share that they will ultimately be answered.

Questions and Answers

QUESTION: How much growth is occurring in the goldfish brain when you are measuring protein synthesis?
ANSWER: The goldfish brain, unlike that of higher animals, continues to grow with age. Some fraction of the incorporation we see may actually represent net increase in brain protein. In the adult mammalian brain, we also see rapid labeling of brain proteins, but this apparently represents only protein turnover. Brain growth in young rats or in the fish does not indicate an increase in the number of neurons since this is fixed early in life. The increased mass is due to increase in other brain elements—myelination, formation of new glial and other supporting cells, and to hypertrophy of existing cells.
QUESTION: How can one distinguish between retrieval and memory loss?

ANSWER: This is a difficult problem. When an animal does not demonstrate a previously learned behavior, we cannot tell whether the behavior is lost or is unretrievable. If, after an apparent loss of memory, we find spontaneous recovery of the learned behavior, it is clear that the observed defect was one of retrieval. In our fish experiments such recovery has not been seen following antibiotic-induced memory loss.

QUESTION: Are there human counterparts of the short- and long-term memory you have observed in the fish?

ANSWER: There may very well be. In rare patients who have bilateral hippocampal lesions, there appears to be a dissociation of short- and long-term memory such as we see in antibiotic-treated animals. Such patients have intact memory of events preceding the brain lesion. They also show acquisition of new information at relatively normal rates, but they cannot retain this new information very long. A similar, though milder, memory deficit is often seen with increasing age in man. Hippocampal lesions in other mammals do not produce comparable behavioral effects.

QUESTION: Do you have an explanation for the environmental effect?

ANSWER: It is possible that the lowering of arousal resulting from the fish being placed back in his home environment triggers the consolidation process. We believe the fish are agitated while in the training situation and do not begin to "fix" memory until they have an indication of success. This message of success could be simply the removal of threat. The survival value of such a triggering mechanism can readily be seen. If an animal's responses serve to remove a threatening environment, he would do well indeed to remember the sequence of actions that removed the threat. While this idea is appealing, it is highly speculative.

Literature Cited

Agranoff, B. W. 1967. Agents that block memory. In *The Neurosciences: A Study Program,* G. C. Quarton, T. Melnechuk, and R. O. Schmitt, eds., pp. 756-764. New York: The Rockefeller University Press.

Agranoff, B. W. 1970. Protein synthesis and memory formation. In *Protein Metabolism of the Nervous System,* A. Lajtha, ed., pp. 533-543. New York: Plenum Press.

Agranoff, B. W. 1971. Effects of antibiotics on long-term memory formation in the goldfish. In *Animal Memory,* W. K. Honig and P. H. R. James eds., pp. 243-258. New York: Academic Press.

Agranoff, B. W., and R. E. Davis. 1968. Evidence for stages in the development of memory. In *Physiological and Biochemical Aspects of Nervous Integration,* F. D. Carlson, ed., pp. 309-327. Englewood Cliffs, New Jersey: Prentice-Hall.

Agranoff, B. W., R. E. Davis, L. Casola, and R. Lim. 1967. Actinomyein D blocks formation of memory of shock-avoidance in goldfish. Science, *158:* 1600-1601.

Agranoff, B. W., and P. D. Klinger. 1964. Puromycin effect on memory fixation in the goldfish. Science, *146:* 952-953.

Appleman, M. M., and R. G. Kemp. 1966. Puromycin: a potent metabolic effect independent of protein synthesis. Biochem. Biophys. Res. Comm., *24:* 564-568.

Casola, L., R. Lim, R. E. Davis, and B. W. Agranoff. 1968. Behavioral and biochemical effects of intracranial injection of cytosine arabinoside in goldfish. Proc. Natl. Acad. Sci., U.S., *60:* 1389-1395.

Davis, R. E., and B. W. Agranoff. 1966. Stages of memory formation in goldfish: Evidence for an environmental trigger. Proc. Natl. Acad. Sci., U.S., *55:* 555-559.

Flexner, L. B., J. B. Flexner, and R. B. Roberts. 1967. Memory in mice analyzed with antibiotics. Science, *155:* 1377-1383.

Glassman, E. 1969. The biochemistry of learning: An evaluation of the role of RNA and protein. Ann. Rev. Biochem., *38:* 605-646.

Hydén, H., and P. W. Lange. 1970. Brain-cell protein synthesis specifically related to learning. Proc. Natl. Acad. Sci., U.S., *65:* 898-904.

Lajtha, A., and J. Toth. 1966. Instability of cerebral proteins. Biochem. Biophys. Res. Comm., *23:* 294-298.

Neale, J. H., and I. Gray. 1971. Protein synthesis and retention of a conditioned response in rainbow trout as affected by temperature reduction. Brain Res., *26:* 159-168.

Yarmolinsky, M. B., and G. L. De la Haba. 1959. Inhibition by puromycin of amino acid incorporation into protein. Proc. Natl. Acad. Sci., U.S., *45:* 1721-1729.

Some Factors in the Regualtion of the Brain's Neurotransmitter Biosynthetic Enzymes and Receptor Sensitivity, Drug Mechanisms, and Behavior

Arnold J. Mandell, David S. Segal,
Ronald T. Kuczenski, and Suzanne Knapp
Department of Psychiatry
University of California at San Diego
La Jolla, California

It has frequently been the case that discoveries have followed a predictable pattern in biology: anatomy followed by physiology; metabolic pathways followed by the mechanisms of their regulation. In a quite similar way, following quickly upon the elucidation of the probable pathways for the biosynthesis of the catecholamines in the adrenal medulla, U.S. von Euler and his group (Bygdeman and von Euler, 1958) as well as others reported interesting relationships between the amount and pattern of medullary hormone release by stimulation and the rate of replenishment by synthesis. This kind of gland-activation sensitive synthetic rate for the adrenal medulla was first reported by Hokfelt and McLean (1950) who showed that stimulation for eighteen periods of one minute during one hour led to an apparent increase in "adrenaline" of 250 percent. This sort of fast adaptive increase in catecholamine biosynthesis during splanchnic nerve stimulation was reported also by Holland and Schumann (1956). An apparently different kind of adaptation requiring more massive stimulation (large doses of insulin, nicotine, or acetylcholine) was reported by West (1951), Udenfriend, et al. (1953), and Butterworth

11

and Mann (1957) who reported an initial period of hours to days of complete "depletion" of adrenal medullary hormone as measured in the effluent which would recover in up to a week with an apparent delayed marked increase above control levels with a gradual decrease to control levels. This and other similar compensatory increases have been interpreted as resulting from the loss of product-feedback inhibition of the rate-limiting enzyme, tyrosine hydroxylase (first demonstrated *in vitro* by Levitt et al., 1965). Aleviation of end-product inhibition could explain the rapid increase in rate of catecholamine biosynthesis caused by a myriad of modulators of sympathetic nervous activity. However, the mechanism of delayed increases preceded by decreases induced by massive stimulation is not as apparent.

In studies of the effects of nerve stimulation and drug manipulation on the isolated guinea pig hypogastric and nerve-vas deferens preparations, Weiner and his group (Thoa et al., 1971) demonstrated two families of adaptive increases: first, a short latency increase in norepinephrine biosynthesis which *was* sensitive to norepinephrine and second, a slower increase which was not. The early change was interpreted as reflecting an alteration in product-feedback inhibition and the later change as reflecting an increase in amount of the rate-limiting enzyme, tyrosine hydroxylase, or substrate availability. Dairman and Udenfriend (1970) induced an increase in sympathetic nerve activity with the administration of an alpha-adrenergic blocking agent and demonstrated a short latency increase in NE biosynthesis; while this short-term increase which was sensitive to ganglionic blocking agents but was not associated with an increase in adrenal or cardiac tyrosine hydroxylase, a slower, long-term increase in NE synthesis was.

A recent series of studies by Mueller, Thoenen, and Axelrod (1969) demonstrated that a number of chronic drug treatments led to increases in the specific activity of tyrosine hydroxylase in the adrenal and sympathetic ganglia which were sensitive to cycloheximide and actinomycin-D. These increases were also inhibited by ganglionic blocking agents and were interpreted as triggered by transynaptic events. Definitive proof that new enzyme protein synthesis is responsible for these kinds of increases in enzyme activity awaits studies using antienzyme antibody and the incorporation of labeled leucine. The inhibitors of macromolecular biosynthesis have multiple actions so that reduction in measurable activity might be due to disruption of the physical state of the enzyme, alteration of substrate supply, and/ or depression of relevant neural function (to name a few factors). The argument made by Thoenen et al. (1971) that the Km of tyrosine hydroxylase activity does not change as an indication that the new activity is due to the new enzyme protein fails to take into account the possibility of a Vmax activation (which will be demonstrated for tyrosine hydroxylase in a later section). The other line of evidence is supplied by studies of mixing various enzyme preparations from treated and untreated animals in various proportions looking for activators and inhibitors. This procedure also is not without serious doubts in that a search for a regulatory physical alteration in enzyme state may require *in vivo* rather than *in vitro* tests (see results section I-A for an example ....). It therefore seems to us that at the outset, we should acknowledge that the discrimination of alterations in states of activation from changes in enzyme amounts in all such studies are not possible definitively by present techniques.

It should be noted that there have been only two previous reports of induced alterations in tyrosine hydroxylase. The first by Musacchio and others (1969) reports increases in brain tyrosine hydroxylase following EST, which was not replicable in our laboratory in several experiments using widely varying parameters and a number of brain regions and subcellular fractions. As a matter of fact, one of the most consistent changes was not an increase, but rather a temporary but statistically significant decrease in enzyme (which reversed in 6 to 8 hours). In contrast, we *have* duplicated the findings reported by Thoenen (1970) demonstrating an increase in brain tyrosine hydroxylase following cold exposure. This work will not be reported here because we are focusing on enzymatic adaptation to behavioral manipulations and psychotropic drugs.

The importance of biosynthetic events in the regulation of synaptically active neurotransmitter has received relatively little attention when compared to neurotransmitter release, reuptake, and storage. This has been particularly true in neuropharmacological models explaining drug action. Specifically, the actions of most of the psychotropic drugs have been explained by their immediate and primary effects on biogenic amine release and uptake processes. It has been only recently that various investigators (Sedvall, Weise, and Kopin, 1968; Kopin et al., 1968; Besson, Cheramy, and Glowinski, 1971) have demonstrated that in active neural systems, functional neurotransmitter is probably newly synthesized transmitter. It is in light of this new data that the previously noted trend in studies of metabolic regulation involving neurotransmitters is regarded as having functional significance.

The neurotransmitter biosynthetic enzymes studied in this work include the initial step branch path enzyme in the biosynthesis of dopamine and norepinephrine, tyrosine hydroxylase; tryptophan hydroxylase, a similarly placed enzyme in the biosynthesis of serotonin; the major biosynthetic enzyme in the synthesis of acetylcholine, choline acetyltransferase; and DOPA decarboxylase. On occasion, biogenic amine degredative enzymes such as catechol-O-methyl-transferase and monoamine oxidase have also been examined. In addition, the specific activity of the various enzymes from regional and subcellular brain fractions following pharmacological and nonpharmacological manipulations, kinetic properties and their physical state characteristics were determined. The specific techniques for these studies are described in the following sections.

It is the purpose of this paper to: (1) summarize some of the work in which we are engaged to elucidate a number of regulatory mechanisms involving the brain's neurotransmitter biosynthetic enzymes; (2) describe what appears to be some aspects of the functional organization of these adaptive mechanisms in relationship to behavior: (3) discuss the implications of these findings for a new neuropharmacological theory of affect disorder.

Materials and Methods

Subjects

Male Sprague-Dawley rats (randomly bred) of the same approximate age and weight for control and experimental groups (130-225 gm) were obtained from Carworth, Inc., Portage, Michigan, and used in most of these studies. Some kinetic studies of tyrosine hydroxylase used the hypothalamic and thalamic regions from 20 kg dogs obtained from the UCSD Department of Animal Resources. Genetic studies of behavior and neurotransmitter enzyme activity were made on five strains of inbred male rats (ACI, BN, BUF, F344, and LEW; generations inbred: F85, F27, F59, F18, and F 59, respectively) obtained from Microbiological Associates, Inc., Walkersville, Maryland. In addition, several studies were carried out using two-week-old White Leghorn chicks from a local supplier.

Methods used with tyrosine hydroxylase

A standardized region of rat midbrain or striate cortex (including a substantial amount of the caudate, the putament, and globus pallidus together weighing approximately 0.03 gm from each hemisphere) was routinely dissected free bilaterally immediately after the sacrifice of the experimental animals. The "caudate" pairs were homogenized in 0.32 M sucrose, weight to volume of 1 to 25, using a Thomas glass-Teflon homogenizer with a 0.025 cm clearance ("wide clearance") using 10 strokes over 2 minutes at 2,000 rpm. The homogenization conditions were standardized as much as possible for all experiments. Samples were then centrifuged in a Sorvall RC-2B at 1,000 x G for 10 minutes, and the low-speed pellet containing nuclear material and cellular debris was discarded. This fraction, which contained 10 to 20 percent of the region's total tyrosine hydroxylase activity, was arbitrarily excluded due to its structural heterogenity. The supernatant was then centrifuged at 11,000 x G for 20 minutes. The supernatant and washed pellet fractions from this separation were used as the enzyme sources in some experiments. In other experiments, the 11,000 x G pellet was washed twice with 3 cc of 0.32 M sucrose followed each time by a 11,000 x G spin for 20 minutes. The pellet was then suspended in 5 cc of 0.32 M sucrose and layered on discontinuous sucrose gradients consisting of 15 cc of 0.8 M and 7.5 cc each of 1.2 M and 1.4 M sucrose. These gradients were spun at 50,000 x G in a SW-25 swinging bucket rotor in a Beckman L2-65B ultracentrifuge for two hours. The bands, corresponding to the P_2A, P_2B, and P_2C fractions of Gray and Whittaker (1962) were collected with a Pasteur pipette. Each band was pelleted by centrifugation in a Ti-60 fixed-angle rotor at 94,000 x G for 15 minutes. Each band was then suspended in 0.32 M sucrose and served as an enzyme source for the appropriate experiments. When enzyme free of subcellular organelles was desired, a 25 to 50 percent ammonium sulfate precipitate of pooled caudate homogenate was used; this precipitate was dialyzed for 12 hours against 100 volumes of 0.002 M potassium phosphate buffer, pH 7.0 before use.

Tyrosine hydroxylase assays were done using a modification of the method described by Nagatsu, Levitt, and Udenfriend (1964). L-Tyrosine-3,5-^{3}H, 39 mC/mmole was purchased from New England Nuclear and purified on a Dowex 50W-X4 (H^+) column; 6,7-dimethyl-5,6,7,8-tetrahydropterine (DMPH$_4$) was purchased from Calbiochem. The incubation mixture contained 1 mμ mole of tyrosine-3,5-^{3}H (containing 10^6 CPM), 380 mμ moles DMPH$_4$, 20 μ moles of mercaptoethanol, 2.8 μ moles FeSO$_4$, and 50 μ moles of sodium acetate buffer, pH 6.1; 50 μl or 100 μl of the enzyme source was added. This mixture was brought to a final volume of 500 μl and incubated for 20 minutes at 37°. The reaction was stopped with 50 μl of glacial acetic acid, and 300 μl of this mixture was placed on a 5.0 cm Dowex 50W-X4 protonated column. The column was washed twice with 400 μl of water. The column eluate and washings were mixed with scintillation fluid and 50 μl of saturated ascorbic acid and counted in a Beckman LS-250 liquid scintillation spectrometer using external standard quench correction. The scintillation fluid contained in 1 liter 150 ml of Biosolv BBS-3 (Beckman) and 850 ml of toluene contained

3.4 gms 2,5-diphenyl-oxazol (PPO) and 0.41 gm 1,4-bis[2-(5-phenyl-oxazolyl)] benzene (POPOP). A rat brain homogenate pool served as the enzyme source of reference for each group of assays. Enzyme activity was linear for time and protein concentration within the parameters of the assay. Dialysis against 100 volumes of 0.32 M sucrose or 0.002 M potassium phosphate buffer, pH 7.4 for 12 hours did not increase enzyme activity in any of the brain-tissue fractions, suggesting that unlike the case with adrenal tissue, catecholamines need not be routinely removed before assaying brain tissue. Protein concentrations were determined using the method of Lowry et al. (1951). Statistical tests of significance were carried out using the Mann-Whitney-U (Siegel, 1956).

Methamphetamine hydrochloride was obtained from Burroughs and Welcome; α-methyltyrosine from Regis Chemical; imipramine hydrochloride from Geigy; reserpine from CIBA; and dl-norepinephrine hydrochloride from Calbiochem.

Methods used with tryptophan hydroxylase

L-Tryptophan-1-[14]C (12 μC/μmole) and 1-tryptophan-3-[14]C (29 μC/μmole) were obtained from New England Nuclear Corp. DL-Dihydroxypehnylalanine-1-[14]C (53 μC/μmole) DOPA) was obtained from Radiochemical Center, Amersham. Impurities were removed from the isotopes by lypholization from 0.1 M Tris-acetate buffer (pH 8.1) or partition chromatography on a Sephedex G-25 column using a 1-butanol-acetic acid-water (20:1:16) solvent for elution. Calbiochem was the supplier of the 6,7-dimethyl-5,6,7,8-tetrahydropterine (DMPH$_4$) and pyridoxal-5[1]-phosphate. DL-Parachlorophenylalanine (PCPA) was obtained from Sigma Chemical Company and Cyclohexamide from Nutritional Biochemicals. Other chemicals were purchased from standard sources in maximal obtainable purity. Rats used in these studies were from a Sprague-Dawley strain (males) obtained from Hilltop (Chicago). They weighed from 130 to 150 grams when studied.

Standardized regions of rat brain (locus ceruleus, median raphe, midbrain tegmentum, corpora quadragemina, substantia nigra, hypothalamus, striate cortex, and septum and frontal cortex, as well as grosser areas such as medulla, pons, midbrain, cerebellum, and the lumbosacral area of the spinal cord) were dissected free immediately after the sacrifice of the experimental animals. Landmarks for these areas were established using a standard rat atlas. Following dissection, the brain tissue was homogenized in 0.32 M sucrose, weight to volume of 1 to 25, with a Thomas glass-Teflon homogenizer with a 0.025 cm clearance using 10 strokes over 2 minutes at 2,000 rpm. Samples were then centrifuged in a Sorvall RC-2B at 1,000 x G for 10 minutes and the low-speed pellet containing nuclear material and cellular debris (and about 20% of the total measurable tryptophan hydroxylase activity) was discarded. Various subcellular fractions were obtained as enzyme sources, including the 12,000 x G pellet ("crude mitochondrial pellet") and supernatant fractions, the 100,000 x G supernatant and pellet fractions ("soluble" and "particulate"), and fractions obtained from discontinuous sucrose gradients. These latter fractions were obtained by suspending the washed 12,000 x G pellets in 5 cc of 0.32 M sucrose and layering them on discontinuous sucrose gradients consisting of 15 cc of 0.8 M, and 7.5 cc each of 1.2 M and 1.4 M sucrose. The gradients were spun at 50,000 x G in a SW-25 swinging bucket rotor in a Beckman L2-65B ultracentrifuge for two hours. The bands, corresponding to the P_2A, P_2B, P_2C, and P_3 fractions of Gray and Whittaker (1962) were collected with a Pasteur pipette. Each band was pelleted by centrifugation in a Ti-60 fixed-angle rotor at 94,000 x G for 15 minutes. Each band was then suspended in 0.32 M sucrose and served as an enzyme source for the appropriate experiments. In some experiments, the enzyme source was the 50,000 x G supernatant fraction, following homogenization in 0:002 M sodium phosphate buffer (pH 7.0) which was called the "shocked soluble" fraction.

Tryptophan hydroxylase assays were carried out using modifications of the method of Ichiyama, Nishizuka, and Hayaish (1970) which in principle couples the hydroxylase with an aromatic amino acid decarboxylase. The two orders of magnitude difference in Km's for tryptophan ($1x10^{-3}$) and 5-hydroxytryptophan ($2x10^{-5}$) of aromatic amino acid decarboxylase make this assay possible. The incubation mixture in the assays of the "soluble" fractions contained 40 μ moles of Tris-acetate buffer (pH 8.1), 380 mμ moles of $DMPH_4$, 10 mμ moles of β-mercaptoethanol, 5 to 10 units of rat kidney aromatic 1-amino acid decarboxylase, 100-200 μl of enzyme preparation (containing 0.3 to 0.6 mg of protein, and 406 mμ moles of 1-tryptophan-1-^{14}C (12 μC/μmoles). These components are reported as optimized. The addition of pyridoxal phosphate did not increase measurable activity. Rat kidney decarboxylase was prepared by homogenization in 0.02 M Tris-acetate buffer, pH 8.1 (1 to 4 weight to volume) followed by a 35 to 50 percent ammonium sulfate precipitation fractionation of the 12,000 x G supernatant fraction which was dialyzed overnight against 100 volumes of 0.002 M sodium phosphate buffer (pH 7.5) before use. The incubation mixture in the assays of the "particulate" fractions contained 40 μmoles of Tris-acetate buffer (pH 8.1), 4 to 6 mμ moles of 1-tryptophan-1-^{14}C (12 μC/μmoles), and 100 to 200 μl of enzyme preparation (containing 0.1 to 0.3 mg of protein) in a final volume of 600 μl. In contrast to assays of the "soluble" fraction, the addition of DMPH$_4$, β-mercaptoethanol or exogenous decarboxylase did not increase the apparent activity (see further discussion of this issue in the results section).

In both the "soluble" and "particulate" assays, before incubation, the reaction mixtures were sealed in 15 cc tubes with rubber caps from which were suspended plastic wells containing 100 μl of NCS (Nuclear-Chicago) for the collection of $^{14}CO_2$. These assays were incubated with shaking for 60 minutes at 37°C. The reaction was stopped with the addition of 500 μl of 2 N perchloric acid injected through the rubber cap with an automatic syringe, following which the $^{14}CO_2$ was evolved by incubation at 37° for three hours in a metabolic shaker. The wells were then removed and placed directly into counting vials containing 10 cc of scintillation fluid consisting of a mixture of toluene fluor (850 cc of toluene containing 3.4 gms 2,5-diphenyloxazol (PPO) and 0.41 gm 1,4-bis [2-(5-phenyloxazolyl)] benzene (POPOP) and 95 percent ethanol in a 1 to 4 ratio. The samples were counted in Beckman LS-250 liquid scintallation spectrometer using external standard quench correction. For both "soluble" and "particulate" assays, an acid or heat denatured enzyme blank was used to determine net activity. Both tryptophan hydroxylase assays were linear with time up to 75 minutes and with protein within the range of protein concentration assayed. PCPA, a known inhibitor of tryptophan hydroxylase (Jequier, Levenberg, and Sjoerdmsa, 1967), inhibited both assays in concentrations of 5 x 10^{-5} to 1 x 10^{-4} M.

The DOPA decarboxylase assay was carried out using a modification of the method of Christian, Diarman, and Undefriend (1970). The incubation mixture contained 40 μmoles of sodium phosphate buffer (pH 6.8), 30 mμ moles of pyridoxal-5-phosphate, 30 mμ moles of β-mercaptoethanol, 1.5 μmoles of 1-^{14}C 1-DOPA (0.5 μC/μM), and 100 to 200 μl of enzyme preparation (containing from 0.1 to 0.2 mg protein) up to a total volume of 400 μl. The mixture was incubated with gentle shaking for 30 minutes at 37°C, after being sealed with rubber caps in the previous assay. The reaction was stopped, the $^{14}CO_2$ evolved and trapped, and the samples counted as previously described. This assay was linear with time up to 40 minutes and protein within the range noted. Protein was determined for specific activity calculation by the method of Lowry and others (1951). Product analysis was used to discriminate hydroxylated from nonhydroxylated, decarboxylated metabolites of tryptophan, as necessitated by the fact that the tryptophan hydroxylase activity was determined by coupling with aromatic amino acid decarboxylase. Products were obtained for chromatography by assays using pargyline hydrochloride (2 μmoles/ assay) and 1-tryptophan-3-^{14}C as substrate.

After stopping the reaction with acidification as described, the reaction mixture was centrifuged at 12,000 x G for 20 minutes and 50 μl of the supernatant along with 2 μg each of 1-tryptophan, tryptamine, 1-5-hydroxytryptophan, and serotonin were applied to 6 x 12 cm silica-G thin layer commercially prepared chromatographic plates and run at constant temperature, unidimensionally, in a closed, presaturated four-phase chromatographic system made up of N-propanol-methyl-acetate-water-ammonium hydroxide, 45-35-10-10. Good separation of the indole amino acids and amines was obtained. Following chromatography and drying, the plates were sprayed with p-dimethylaminobenzaldehyde 1 percent in acidified ethanol. Indole spots were visualized within five minutes and were scraped from the plates and counted in scintillation vials containing 10 cc of counting fluid as described. Background counts were obtained from equal areas of silica plate collected from between the visualized indole spots, and up to 80 to 85 percent recovery of the total radioactivity applied to each plate was recovered as indole amino acids or amines.

Methods used with choline acetyltransferase

Two-week-old White Leghorn chicks were injected intraperitoneally with various drugs and the optic lobes (chosen due to their high levels of choline acetylase and clear-cut anatomical demarcation) were removed and homogenized 1 to 4 (weight to volume) in 5 mM NaPO$_4$ buffer (pH 7.5) at 4°C. Assays for optic lobe choline acetyltransferase were carried out using the radiochemical method of Schrier and Shuster (1967). Protein determinations were done using the method of Lowry and others (1951). Five to seven animals were studied in each group.

Techniques used for measurement of rat behavior

The spontaneous activity of each rat was measured in a soundproof chamber (18″ x 18″ x 11″). The floor of the chamber was divided into quadrants and an index of gross activity was obtained from the number of crossovers each animal made from one quadrant to another. In addition, rearings were measured by the use of touch-plates placed 5½ inches above the floor (Segal and Mandell, 1970).

To obtain baseline levels of spontaneous activity, each animal was habituated to the activity box before experimental observation. On the following experimental days, the animals were exposed to the apparatus for varying periods of time dictated by the experimental design.

Electronmicroscopy of subcellular fractions

Samples of the washed and resedimented subcellular fractions were fixed for three hours in 2 percent O_8O_4 in Veranol buffer (pH 7.5). The pellets were subsequently dehydrated through a graded series of ethanol solutions, embedded in an epoxy resin mixture, and allowed to polymerize for 48 hours at 60°C. Thin sections (approximately 600 millimicrons) were cut with a diamond knife on a LKB Ultramicrotome. These sections were examined and electronmicrographs were made using a Zeiss 9S Electromicroscope. Prints examined ranged from 18,000 to 32,000 x magnification.

Results

Apparent regulatory mechanisms affecting neurotransmitter biosynthetic enzymes or receptors

Drug induced change in the physical state of tyrosine hydroxylase. The administration of methamphetamine intraperitoneally to rats in doses from 1 to 5 mg/kg led to a dose-related shift in the subcellular distribution of caudate tyrosine hydroxylase activity. A significant amount of enzyme activity appeared to shift from the 11,000 XG supernatant fraction to the p_2 B and p_2C (Gray and Whittaker, 1962) particulate fractions with no change in total measurable activity (Fig. 1). Using the "crude mitochondrial pellet" for a determination of the time course of this event, Figure 2 demonstrates a shift of tyrosine hydroxylase activity from the "soluble" to the "particulate" fractions following a dose of 5 mg/kg of methamphetamine which was maximal in 30 minutes and lasted six hours. As can be seen, there was no change in total measurable activity at any time. As mentioned, this was a somewhat dose-related phenomenon, both in latency and duration. Figure 3 demonstrates a 10-minute latency for full effect at 15 mg/kg of methamphetamine and a moderate increase in duration (from 6 to about 8 hours).

This effect was *not* obtainable with *in vitro* additions of methamphetamine or catecholamines during homogenization. The administration of impramine hydrochloride, intraventricular norepinephrine infusions, electroconvulsive shock, or intermittent footshock also failed to lead to this phenomenon. As seen in Figure 4, α-methyltyrosine (150 mg/kg) administered one hour before saline or methamphetamine, 5 mg/kg, to animals sacrificed one hour later produced a similar and possibly additive effect. We are currently speculative that this short-latency, reversible shift in the physical state of striatal tyrosine hydroxylase may be triggered by the acute reduction in intraneuronal catecholamine; perhaps caudate dopamine.

THE EFFECT OF AMPHETAMINE ON THE SUBCELLULAR DISTRIBUTION OF TYROSINE HYDROXYLASE*

	% OF TOTAL ACTIVITY	
	SALINE	AMPHETAMINE
11,000 x G SUPERNATENT	69	36
P_2A	0	0
P_2B	16	35
P_2C	14	27

TOTAL ACTIVITY
$\mu\mu$M DOPA/mg prot/hour

SALINE   102.3 $\pm$ 19.4

AMPHETAMINE   81.0 $\pm$ 16.7

*Four hours after 15 mg/Kg; means of
three pools of three caudates.

FIGURE 1.   The total activity and subcellular distribution of rat striatal tyrosine hydroxylase studied two hours after the administration of methamphetamine hydrochloride, 1 mg/kg subcutaneously or an equivolume amount of saline. The values equal the means of three pools of three caudate areas each. The subcellular distribution is expressed as the mean percent of total measurable activity contained in each fraction. Amphetamine produced no effect on total activity, which is expressed as $\mu\mu$ moles DOPA synthesized per hour. Note the shift of enzyme activity from the soluble to the "synaptosomal" and "mitochondrial" fractions.

Studies characterizing the particulate state of tyrosine hydroxylase. Wurzburger and Mussachio (1971) have recently reported a tendency of otherwise soluble adrenal tyrosine hydroxylase to form nonspecific aggregates during sucrose homogenization. In contrast to the adrenal enzyme, a substantial fraction of striatal tyrosine hydroxylase remains particulate not only during sucrose homogenization but also in hypotonic (0.002M potassium phosphate buffer, pH 7.9) media. In addition conditions which tend to disrupt synaptosomes such as narrow clearance (0.009 cm) homogenization and sonication (Hillarp, 1960) do not alter the particulate nature of tyrosine hydroxy-

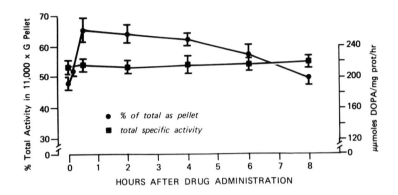

FIGURE 2.   The total measurable striatal tyrosine hydroxylase activity ($\mu\mu$ moles DOPA synthesized per hour) and the percent of that total in the post low-speed spin, 11,000 x G pellet at various times after methamphetamine hydrochloride, 5 mg/kg. Each point represents the mean and the standard error of the mean of six caudate pairs (six animals). Whereas total measurable activity was not changed by the administration of methamphetamine, the percent in the pellet rose rapidly, reaching its peak in 30 minutes, and remained significantly different from saline controls (which were not affected) until six hours.

lase. Our laboratory has recently carried out experiments in which 100,000 x G supernatant (soluble) midbrain tyrosine hydroxylase was mixed with a number of subcellular fractions from rat forebrain prepared by Cotman (Cotman and Flansburg, 1970). The enzyme exhibited preferential binding to the synaposolam membrane fraction (in contrast to the plasma membrane, mitochondrial, and other fractions). The suggestion of nerve-ending binding specificity for brain tyrosine hydroxylase is also substantiated by experiments in which brain subcellular fractions and 100,000 x G soluble tyrosine hydroxylase were incubated with and without Ca^{++}. The fractions with the most native tyrosine hydroxylase bound the most additional tyrosine by hydroxylase, and the binding was promoted by Ca^{++}. We have, therefore, tended to regard the drug alterable particulate state of striatal tyrosine hydroxylase as a potential physiological mechanism in which the enzyme can be bound to specific nerve-ending membranes and in which intraneuronal catecholamine concentration and/or Ca^{++} may play a role.

Studies relating the particulate state of tyrosine hydroxylase to decreased specific activity. Membrane binding as a mechanism for regulating enzymes has been well described for mitochondrial en-

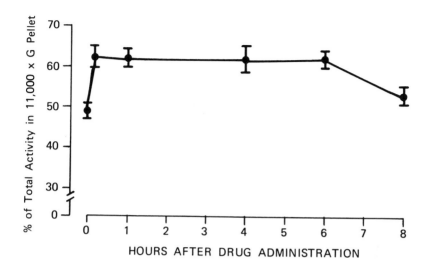

FIGURE 3.   The latency and duration of the methamphetamine-induced relative increase in particulate tyrosine hydroxylase at various times following the highest administered dose of the drug (15 mg/kg). Each point represents the mean and the standard error of the mean for six pairs of caudate (six animals) of the percent of total striatal tyrosine hydroxylase in the previously defined 11,000 x G pellet. Note that the effect appeared to reach its peak in ten minutes and returned to control levels at eight hours.

zymes such as succinate dehydrogenase and other flavoprotein enzymes (Porcellati and diJeso, 1971). This kind of mechanism has not been described previously for neurotransmitter enzymes in the brain. By manipulating ion concentration, Ca^{++} concentration, subcellular fraction combinations, and the addition of sulfated mucopolysaccharides, we have created a number of *in vitro* models demonstrating both decreases and increases in the specific activity of tyrosine hydroxylase in the bound versus the soluble state.

The possibility that *occlusive binding* of striatal tyrosine hydroxylase could occur was demonstrated by our experiments showing that the addition of Ca^{++} to the 3,000 x G supernatant fraction of striatal homogenate markedly decreased the level of measurable tyrosine hydroxylase activity. If this homogenate was then spun at 17,000 x G for 20 minutes and the Ca^{++} removed by dialysis, the latent activity was recovered in the particulate fraction. This was seen as consistent with a Ca^{++} promoted binding phenomenon associated with a reversible inhibition (occlusion). In a similar set of experiments, summarized in Figure 5, 17,000 x G pellets from rat caudates that were hypotonic-ally shocked in 0.002 potassium phosphate were assayed in the standard assay and in an assay that optimized binding using 0.32 M sucrose solution instead of water with a 0.001 M histidine buffer (pH 6.2). When tyrosine hydroxylase obtained from the 17,000 x G supernatant fraction or a 25 to 50 percent ammonium sulfate precipitate (followed by dialysis) was assayed in the sucrose assay, there was a 47 to 50 percent reduction in activity compared to the standard assay. Kuczenski and Mandell have recently demonstrated a relatively nonspecific ion-activation of soluble brain tyrosine hydroxylase with a wide variety of ionic species (1971). It is within this context that the decreased activity of the soluble enzyme in a nonionic media is understood. When the enzyme activities of the shocked 17,000 x G pellets were compared in the two assays, however, the sucrose assay resulted in a 76 ± 2 percent decrease in specific activity. Here also under conditions of increased membrane binding there was a significant inhibition of tyrosine hydroxylase beyond the decreased activity of the enzyme in the nonionic assay alone. Benson, Cheramy, and Glowinski (1971) have recently demonstrated the relatively quick induction of noncompetitive inhibition of dopamine biosynthesis in rat striatal brain slices following pretreatment with amphetamine at 5 mg/kg. It was unlikely that this was the result of an inhibition of the enzyme by

Injection #1	Injection #2	% Shift Over Saline
Saline	Saline	0 $\pm$ 3.5%
Saline	Methamphetamine	11.8 $\pm$ 0.5%
αMT	Saline	11.8 $\pm$ 1.1%
αMT	Methamphetamine	15.4 $\pm$ 2.3%

FIGURE 4. The effect on the size of the methamphetamine-induced difference between the percent of the total measurable caudate tyrosine hydroxylase in the 11,000 x G pellet and the 11,000 x G supernatant fractions of pretreatment with 150 mg/kg of α-methyltyrosine (αMT). Since these fractions usually equal each other (about 50% in the 11,000 x G supernatant and particulate), no drug effect would make the difference equal approximately zero. Each number represents the mean and standard error of the mean of ten caudate pairs. The pretreatment ("Injection #1") preceded "Injection #2" by one hour; the animals were sacrificed one hour after the second injection. The difference between the percent of total activity in the pellet and the supernatant after two saline injections was arbitrarily set at zero and subtracted from the following single or double drug experiments. Note that methamphetamine (as seen previously) increases the relative amount of enzyme activity in the 11,000 x G pellet as does α-methyltyrosine. These effects were partially additive.

amphetamine on tyrosine hydroxylase at concentrations as high as $10^{-2}M$ *in vitro*. It may be that this amphetamine induced inhibition of dopamine biosynthesis in rat striatam is due to the shift of tyrosine hydroxylase to an occlusively membrane bound state.

Studies relating the particulate state of tyrosine hydroxylase to increased specific activity. In addition to the nonspecific ionic effects of such salts as NaCl, K_3PO_4, NH_4Cl, and KCl (Fig. 6), we have found that $SO_4 =$ ion activated soluble brain tyrosine hydroxylase greater than the effect of ionic strength. In addition to sulfate effects, a highly specific activation of tyrosine hydroxylase was observed with the sulfated mucopolysaccharides. Whereas the nonsulfated congeners (hyluronic acid or glycogen) had no effect, chondroitin sulfate C activated the enzyme by 25 percent (Fig. 7). What is even more interesting is that the sulfated mucopolysaccharide, heparin, increased tyrosine hydroxylase activity over 300 percent with a K_a near 0.35 M. The slope of a Hill plot of the data (Koshland, 1970) yielded a value for n_H of about 2.3, suggesting a positive allosteric effect of heparin.

The Effect of an Ionic Medium on the Subcellular Distribution and Specific Activity of Caudate Tyrosine Hydroxylase

	Subcellular Distribution (% total activity)		Specific Activity (ionic assay = 100%)	
	.32 M Sucrose	.001 M Na$_2$PO$_4$	Ionic Assay	Non-ionic Assay
Supernatant	48%	78%	100%	53 ± 4%
Pellet	52%	22%	100%	24 ± 2%

FIGURE 5.   Whereas sucrose homogenization resulted in an approximately 50-50 distribution of tyrosine hydroxylase activity following 17,000 x G centrifugation between supernatant and pellet tyrosine hydroxylase activity, hypotonic homogenization resulted in more enzyme solublized. When these fractions were assayed in the standard assay ("ionic") and in an assay that optimized binding, using 0.32 M sucrose solution with a 0.001 M histidine buffer (pH 6.2), one can see a sucrose-induced reduction of 47 percent but a binding-induced reduction of 76 percent. This has been seen as an *in vitro* model for occlusive binding. (See the text.)

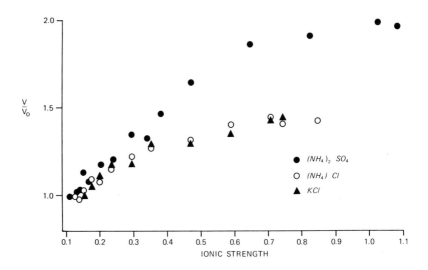

FIGURE 6.   Effect of ionic strength on the activity of dog hypothalamic tyrosine hydroxylase in 0.11 M Tris-acetate buffer, pH 6.1, at 37°. The assay mixture contained 3 x 10^{-6} M tyrosine-3,5-H^3 (containing 10^6 cpm), 1.1 x 10^{-3}H DMPH$_4$, 0.014 M β-mercaptoethanol, and 8.76 x 10^{-6} M FeSO$_4$. Each assay contained 250 μgm protein.

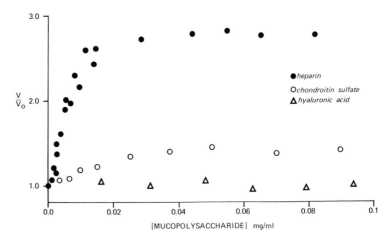

FIGURE 7.   Effect of polysaccharides on the activity of tyrosine hydroxylase. See the legend to Figure 6 for assay conditions.

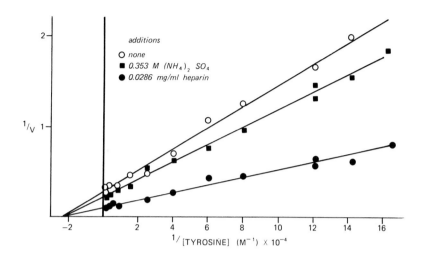

FIGURE 8. Lineweaver-Burke plot of the activity of hypothalamic tyrosine hydroxylase as a function of tyrosine concentration. Assay conditions were as in Figure 6 except for tyrosine.

Whereas both sulfate and heparin failed to alter the Km of the enzyme for substrate (Fig. 8), heparin (0.0286 mg/ml of assay mixture) markedly decreased the Km of the enzyme for its co-factor, $DMPH_4$, as well as increasing the V_{max} of the enzyme twofold (Fig. 9). It is of interest that this kind of allosteric activation of the enzyme by sulfated mucopolysaccharides does not occur when the enzyme is studied in the particulate state or the soluble enzyme is allowed to bind the nerve-ending membranes. In this regard it is of interest that Elam and others (1970) have reported a heparin-like compound rapidly transported along axons to nerve endings and that the subcellular fraction from brain with the highest concentration of sulfated mucopolysaccharides was that containing the synaptic vesicles (Vos and Roberts, 1968). It is therefore possible that heparin-like regions in nerve-ending membranes may serve as sites for allosteric binding activation. Since this activation markedly increases the affinity of the enzyme for its cofactor, $DMPH_4$, and its product-feedback competitors, the catecholamines, it may be that this activation sensitizes the enzyme to alterations in levels of the end product. An idea of the potential regulatory function of this allosteric activation vis-á-vis cofactor and

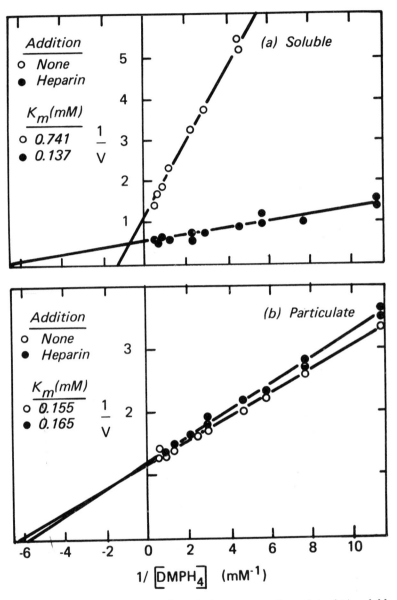

FIGURE 9. Lineweaver-Burke plot of the activity of caudate (A) soluble tyrosine hydroxylase and (B) particulate tyrosine hydroxylase as a function of DMPH₄ concentration in the presence and absence of 0.0286 mg/ml heparin. Assay conditions are as in Figure 6.

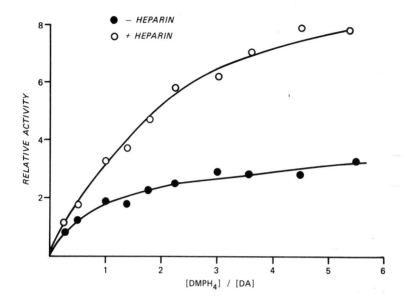

FIGURE 10.   The effect of heparin activation on the relationship between the ratio of cofactor to inhibitor at varying concentrations of each and the relative specific activity of brain-soluble tyrosine hydroxylase. Note that over a rather wide range of this ratio, activation by this sulfated mucopolysaccharide changes the slope of this relationship. This can be seen as turning the biosynthetic mechanism to be more sensitive to changes in either cofactor or end-product concentrations.

product concentrations can be obtained from Figure 10, in which the relative specific activity of the enzyme is plotted as a function of the DMPH₄/DA ratio, with and without sulfated mucopolysaccharide activation. Particularly at low levels of cofactor (or high levels of catecholamine product), the conformational "activation" can be seen as tuning the biosynthetic mechanisms to be more sensitive to both cofactor and product concentrations so that small changes in either one have a larger capacity to alter the activity of the enzyme.

An example of a drug-induced rapid activation of brain particulate tyrosine hydroxylase in vivo.   In some recent studies involving the characterization of enzymatic responses to mood-altering drugs, we happened upon an agent which appeared to rapidly activate predominantly particulate tyrosine hydroxylase from rat striate cortex without altering the predominantly soluble enzyme in the midbrain. The heraadrenergic blocking agent, propranolol (Inderal (R)), used in cardiac anythmias and recently in acute psychoses (Sullivan et al., 1971) pro-

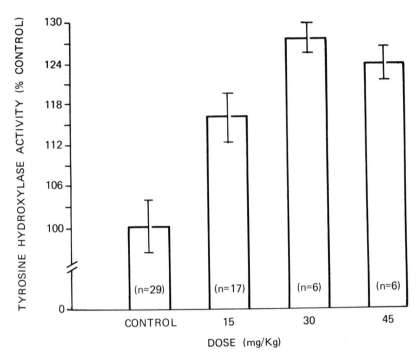

FIGURE 11. The effects of three doses of parenterally administered propranolol on striatal tyrosine hydroxylase activity one hour following drug administration. Enzyme activity is expressed as the percent of saline control values and represents the total measurable activity per "caudate" pair. Both 15 mg/kg and 30 mg/kg led to values significantly higher than control ($p < .05$).

duced a dose-dependent, reversible increase in striatal tyrosine hydroxylase which reached maximal levels in one hour. Figure 11 demonstrates the effect achieved with 30 mg/kg. Figure 12 demonstrates the time course of this effect at 30 mg/kg in which observable changes occur within 30 minutes, are optimal within 60 minutes, and return to control values by four hours. Addition of the drug and its probable metabolites from drug-treated enzyme preparations at various concentrations failed to produce this effect. Pretreatment with the inhibitor of protein synthesis, cycloheximide, at a level of the incorporation of ^{14}C-leucine into brain trichloroacetic acid precipitable protein failed to alter the magnitude of the propranolol response. It is our interpretation of this phenomenon that although the intervening events have not

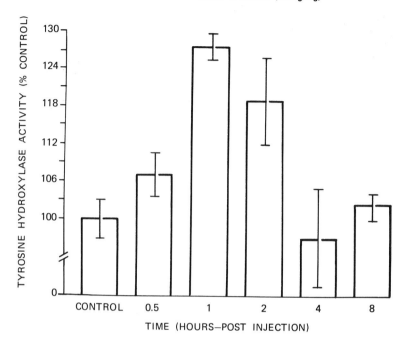

FIGURE 12. Effect of a single dose (ip) of propranolol, 30 mg/kg, on total striatal ("caudate") tyrosine hydroxylase activity at various times after drug administration. Highest values were obtained at the one-hour interval, with return to control values atfour hours. Both the one- and two-hour interval determinations were significantly greater than control (p < .05).

been established (receptor blockade followed by feed-back activation; increased intraneuronal catecholamine secondary to reduction in release by propranolol), represents an example of an alteration or increase in membrane binding associated with allosteric activation similar to that demonstrated *in vitro* with sulfated mucopolysaccharides.

Further studies relating these *in vitro* regulatory membrane-binding models to *in vivo* effects are in progress. Some of the complexity that may be involved, however, in the interpretation of the data we are collecting about *in vitro* tyrosine hydroxylase in relationship to *in vivo* functional state is exemplified by an attempt to explain the significance of the propranolol-induced enzyme changes. For example, one might speculate that the propranolol-induced increase in brain catecholamines results not from receptor blockade but a drug-induced inhibition of biogenic amine release (von Euler and Lishajko, 1966). If this led to an allosteric binding activation, it would sensitize the nerve-ending tyrosine hydroxylase to the negative-feedback effects of the increased intraneuronal catecholamines (competition for the $DMPH_4$ site) so that although we would observe an *increase* in activity in the standard particulate assay (with diluted or dialyzed catecholamines) the functional effects would be decrease in catecholamine synthesis. This same kind of *in vitro* enzymatic change has been observed with an acute and high dose of reserpine, 5 mg/kg, which depletes intraneuronal catecholamines (although the drug's effect on intraneuronal extravesicular catecholamines is indeterminate). Figure 13 is a summary of such an experiment demonstrating that there is a marked (but not reversible) increase in activity in one hour *in vivo*. In this case we would speculate that although allosteric activation of the cofactor site has occurred via the appropriate binding, the depleted intraneuronal catecholamine state would lead to a functional increase in enzyme activity due to the increased affinity for $DMPH_4$, an increase in V_{max} for substrate, and a decrease in the catecholamine level competing for the cofactor site. There appears to be mechanism for "tuning up" the enzyme to both the positive effects of cofactor ($DMPH_4$) or the product inhibitory effects of the appropriate catecholamine. The net effect on synthetic function, therefore, appears dependent on both the alterable physical state as well as the changing levels of intraneuronal neurotransmitter. This mechanism can be seen as an *in vitro* model in Figure 10.

Regulation of neurotransmitter enzyme via regulation of substrate availability. Early studies of the branch pathway initial enzyme in serotonin biosynthesis, tryptophan-5-hydroxylase in mammalian brain (Ichiyama et al., 1970; Grahame-Smith, 1964; Gal, Morgan, and Marshall, 1965) used enzyme in two forms: particulate enzyme from the midbrain from the "crude mitochondrial pellet" prepared from homogenization in 0.32 M sucrose, constituting less than 20 percent of the midbrain enzyme which is 80 percent soluble, or soluble enzyme as the post-spin soluble fraction (50,000 x G; with or without hypotonic lysis). Our studies of tryptophan hydroxylase activity in various brain regions led us to conclude that the particulate form of the enzyme, requiring careful preparation in 0.32 M sucrose to maintain its activity, constituted a significant amount of the enzyme in areas where serotonergic nerve endings were prominent (such as frontal cortex, septal area, caudate, and lumbosacral cord). The soluble form of the enzyme is seen as predominant in the midbrain and

RESERPINE TIME COURSE

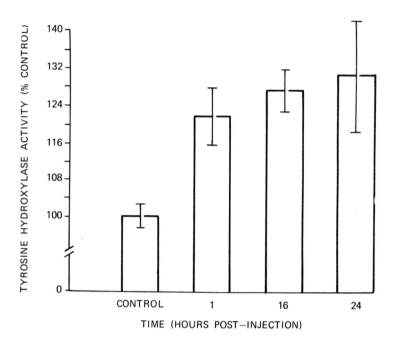

FIGURE 13.   Effect of a single dose (ip) of reserpine, 5 mg/kg, on triatal tyrosine hydroxylase activity at various times following drug administration. There was a significant increase in enzyme activity at one hour which lasted at least 24 hours ($p < .05$); n = 6 for each group. (See the text.)

the pons-medulla area where serotonergic cell bodies have been reported (Hillarp, Fuxe, and Dahlstom, 1966). Figure 14 summarizes the subcellular distribution of tryptophan-5-hydroxylase activity in various brain regions. Electron micrographs of both subcellular fractions and regions with high density of particulate enzyme demonstrated a high density of synaptosomes or nerve endings. These two physical forms of enzyme are distinguishable in a number of ways. In the *in vitro* assays of both forms, whereas the coupled hydroxylase-decarboxylase assay (see Methods and Materials) could be stimulated by the addition of rat kidney decarboxylase and $DMPH_4$, the particulate could not. Whereas the soluble enzyme form resisted dialysis against hypotonic solutions (0.002 potassium phosphate), the particulate form lost activity which was *not* recoverable with the addition of DOPA decarboxylase and $DMPH_4$. That hypotonic effects on the

PHYSICAL STATE OF REGIONAL BRAIN TRYPTOPHAN HYDROXYLASE

BRAIN PART	$\mu\mu$moles/mg protein		$\mu\mu$moles/region		
	soluble	particulate	soluble	particulate	% soluble
WHOLE BRAIN	18.4	10.8	571.8	259.5	68.0
MEDULLA	3.7	19.9	4.4	40.6	9.6
LOCUS CERULEUS	69.0	29.0	13.0	16.0	44.0
PONS	42.9	19.8	60.5	68.7	47.6
MIDBRAIN	87.1	14.4	228.1	75.0	75.0
MEDIAN RAPHE'	412.6	31.0	53.0	13.5	80.0
MESENCEPHALIC TEGMENTUM	99.0	66.0	241.0	208.0	50.0
SUBSTANTIA NIGRA	70.0	57.5	7.0	17.0	27.0
CEREBELLUM	0	4.8	0	11.0	0
HYPOTHALMUS	29.8	29.3	31.3	82.4	28.0
CAUDATE	0	100.6	0	300.0	0
SEPTUM	0	60.5	0	100.0	0
FRONTAL CORTEX	0	41.1	0	60.4	0
LUMBO SACRAL SPINAL CORD	10	26.0	22.0	99.0	20.0

FIGURE 14.   A summary of the specific activity and subcellular distribution of brain tryptophan hydroxylase. Note that areas that are known to have serotonin cell bodies are dominated by soluble enzyme; those which are characterized by serotonin nerve endings evidence mostly particulate enzyme. (See the text.)

second coupled enzyme in the particulate assay (endogenous tryptophan or DOPA decarboxylase) was not responsible for the lost activity with lysis was demonstrated by the manifold *increase* in measurable DOPA decarboxylase activity via lysis. Another treatment which differentially reduced the particulate activity was storage for three or four hours at 37° (which also did not affect the decarboxylase activity). It began to appear to us that either a specific physical relationship between the synaptosomal tryptophan hydroxylase and tryptophan decarboxylase was required, or that a physical arrangement was necessary to maintain a necessary and perhaps regulatory substrate uptake mechanism.

Another example of the differentiability of the two physical forms of this enzyme was their response to the acute and chronic administration of morphine. Figure 15 demonstrates that whereas neither the acute or chronic treatment of rats with morphine produces a change in the midbrain (soluble) enzyme form, the septal (particulate) en-

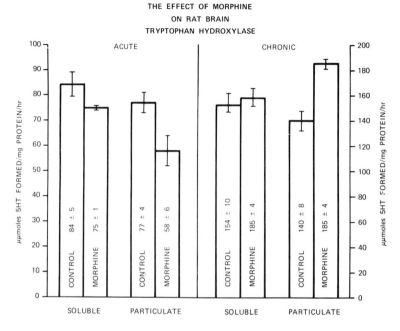

THE EFFECT OF MORPHINE
ON RAT BRAIN
TRYPTOPHAN HYDROXYLASE

FIGURE 15.   The effect of subcutaneously administered morphine sulfate on the specific activity of midbrain (soluble) and septal (particulate) tryptophan hydroxylase. Morphine was administered in the acute experiments at a dose of 10 mg/kg. The chronic experiments consisted of subcutaneous implantation of a morphine sulfate supplied by E. Way. Note that the septal particulate activity dropped acutely (as it did following the administration of a number of agents such as amphetamine and cocaine) and increased in the chronically treated state. N = 6 for each group. (See the text.)

zyme is initially reduced in activity, followed by longer latency increase. In an experiment using parachlorophenylalanine (Fig. 16), a reversible and then irreversible inhibitor of tryptophan-5-hydroxylase *in vivo*, note that the initial effect on particulate enzyme is a quickly reversible decrease in activity. The other findings in this figure will be discussed in a later context. Grahame-Smith (1968) has recently demonstrated using uptake of labeled tryptophan by Millipore-filtered synaptosomes, that PCPA competitively inhibits uptake of label. Whether this uptake process is associated with 5-hydroxylation was not determined. This process was dependent on Ca^{++}, Mg^{++}, or Na^+ but not K^+ and was destroyed by both storage at 37° and hypotonic shock. Thus, Grahame-Smith's studies of the conditions affecting the synaptosomal uptake process of tryptophan substrate and our studies of the variables altering particulate but not soluble enzyme activity appear quite similar. It seems clear that the physical integrity of the nerve-ending serotonin biosynthetic unit as well as certain ions are required for function and that one of the consequences of this arrangement is the regulation of enzyme activity via controlling substrate supply. It is interesting that Grahame-Smith has recently demonstrated an almost linear relationship between brain tryptophan levels and rate of serotonin biosynthesis—this was to be expected in that tryptophan hydroxylase is far from saturated at brain-tissue levels of tryptophan. Recently Perez-Cruet and others (1971) have demonstrated that a drug-induced (lithium) facilitation of this substrate promoted increase in brain serotonin biosynthesis. It thus appears that another

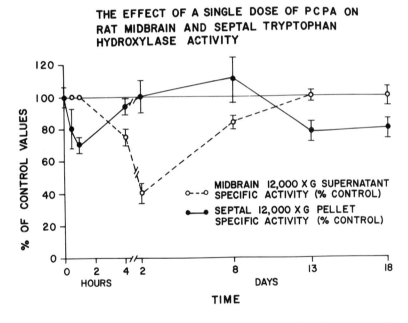

THE EFFECT OF A SINGLE DOSE OF PCPA ON RAT MIDBRAIN AND SEPTAL TRYPTOPHAN HYDROXYLASE ACTIVITY

MIDBRAIN 12,000 X G SUPERNATANT SPECIFIC ACTIVITY (% CONTROL)

SEPTAL 12,000 X G PELLET SPECIFIC ACTIVITY (% CONTROL)

% OF CONTROL VALUES

TIME

HOURS   DAYS

FIGURE 16.   The activity of midbrain and septal tryptophan hydroxylase following the acute administration of PCPA, 300 mg/kg. Note the initial reversible decrease in septal enzyme activity followed by the return to control levels and then a delayed fall. At the same time, midbrain enzyme is decreased more slowly and more profoundly with a delayed return to control levels. These data are seen as exemplifying an acute effect on substrate availability in particulate enzyme followed by an apparent axoplasmic flow from midbrain to septum of defective tryptophan hydroxylase. (See the text.)

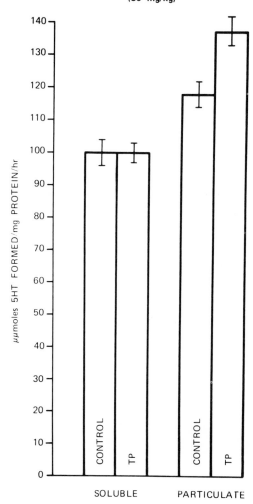

TRYPTOPHAN HYDROXYLASE
3 HRS. AFTER TRYPTOPHAN LOAD
(50 mg/kg)

FIGURE 17.   This graph demonstrates the effect of L-tryptophan load, 50 mg/kg
(ip), on septal-particulate tryptophan hydroxylase. This apparent activation had
a latency of three hours and was apparent at a time when brain tryptophan
would be returned to control levels (Grahame-Smith, 1968). (See the text.)

mechanism available for neurotransmitter biosynthetic enzyme regulation is the physiological or pharmacological alteration of substrate availability.

Regulation of a neurotransmitter enzyme via substrate activation. As noted above, studies such as those by Grahame-Smith (1971) relating tissue tryptophan levels in brain to serotonin levels suggest that synthesis is controlled via substrate availability. Studies of the early effects of PCPA as mentioned above are consistent with this view. This issue has recently seemed more complex in that we have seen that L-tryptophan loads (over 50 mg/kg administered intraperitoneally) leads to a small but significant increase in particulate but not soluble tryptophan hydroxylase activity with a latency of two hours. Figure 17 demonstrates this substrate-induced increase in enzyme ac-

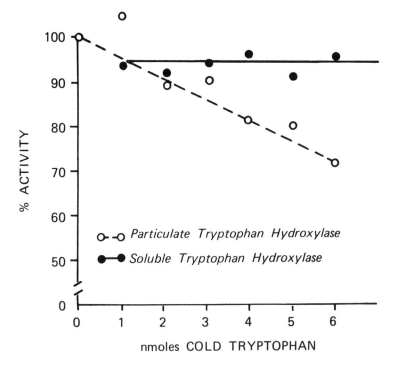

FIGURE 18.   The effect of increasing the nonisotopic L-tryptophan levels beyond that in the standard assay conditions on the specific activity of soluble and particulate tryptophan hydroxylase. Note that there appears to be the expected dilution of isotopic CO_2 product in the substrate uptake-limited particulate fraction and not in the soluble fraction.

tivity. When cold L-tryptophan levels are gradually increased *in vitro*, the manifest level of activity is reduced as expected; the cold trpytophan diluting the labeled substrate at the uptake or enzymatic step (Fig. 18). This interesting phenomenon, reminiscent of the substrate induction of tryptophan pyrollase in the liver (Grungard and Fergelson, 1961) may involve altered kinetics for the various components in the assay. Further exploration of this substrate activation phenomenon is in progress. However, Grahame-Smith (1971) has shown that at the time we reported the enzyme elevation (three hours), the brain tryptophan levels are returned to normal after tryptophan load.

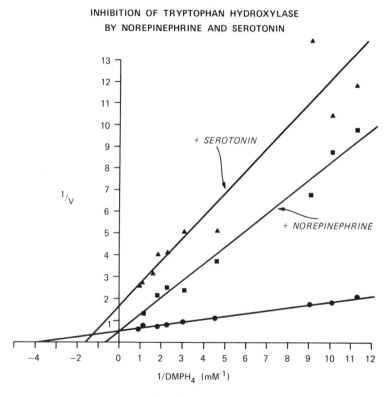

INHIBITION OF TRYPTOPHAN HYDROXYLASE BY NOREPINEPHRINE AND SEROTONIN

FIGURE 19. A Lineweaver-Burke plot of the activity of soluble tryptophan hydroxylase as a function of cofactor concentration under standard assay conditions and with norepinephrine or serotonin added. The Ki for the apparent competitive inhibition of the enzyme by norepinephrine is 1.7×10^{-4} M at a norepinephrine concentration of 5×10^{-4} M. Serotonin appears to lead to a noncompetitive inhibition at a relatively high concentration (1×10^{-3} M).

Regulation of a neurotransmitter enzyme via competitive inhibition by a product of another neurotransmitter biosynthetic pathway. In our studies establishing the kinetic characteristics of the soluble and particulate tryptophan hydroxylase, it was logical to determine whether this enzyme *in vitro* had the same sensitivity to product feedback inhibition as did tyrosine hydroxylase. Figure 19 shows that soluble enzyme manifests a Km for $DMPH_4$ of 3.2 x 10^{-4} at optimal concentrations of substrate and rat kidney decarboxylase. Figure 19 also demonstrates a competitive inhibition by norepinephrine for the cofactor, $DMPH_4$, similar to that reported for tyrosine hydroxylase. The Ki for this competitive inhibition was calculated to be 1.7 x 10^{-4} M at a norepinephrine concentration of 5 x 10^{-4} M. The branch pathway product, serotonin, appears to noncompetitively inhibit the $DMPH_4$ - tryptophan-5-hydroxylase interaction at a concentration of 1 x 10^{-3} M. Due to the tendency for indoles and indoleamines to nonspecifically bind to protein, it is likely that this is a nonspecific effect. This presents an interesting situation in which a catecholamine is a far more effective inhibitor of an indoleamine biosynthetic pathway than its indoleamine product.

Neurotransmitter biosynthetic enzyme regulation by long lasting increases in specific activity (new enzyme synthesis?). Stimulated by the results of a number of workers studying enzymatic adaptation in peripheral adrenergic systems (Dairman and Udenfriend, 1970; Mueller, Thoenen, and Axelrod, 1969; Thoenen et al., 1971; Weiner, 1970; Mandell and Morgan, 1970), we have been using various pharmacological, physiological, and behavioral manipulations to alter the brain's neurotransmitter biosynthetic enzymes. The general strategy has been the use of a treatment for several days before determining enzyme-specific activity (Segal et al., 1971). In the case of drugs, the administered dose is lower and given regularly so that blood and brain levels of the agent are maintained over time. Time and dose parameters in this kind of experiment are critical. For example, increases in septal particulate tryptophan hydroxylase following several days of treatment with morphine, do not occur unless the doses are given close enough together to maintain the tissue level or a pellet is implanted under the skin which will release morphine continuously. When these new states were chronically induced, usually the specific activity of the neurotransmitter biosynthetic enzyme changed gradually, reached a peak (or trough) in several days, and remained at this new level for several more days. Due to the chronic nature of these experiments and the toxicity of inhibitors of protein or nucleic acid synthesis, it has not been possible to establish whether these changes are due to alterations in *amount* rather than activity of the enzymes, although the

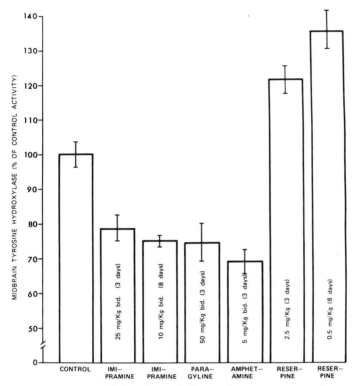

FIGURE 20. A summary of the effects of chronic drug treatment on the activity of midbrain tyrosine activity. Reserpine given in daily doses of 2.5 mg/kg for three days, 0.5 mg/kg for eight days; imipramine, 25 mg/kg b.i.d. for three days, 10 mg/kg b.i.d. for eight days; pargyline, 50 mg/kg b.i.d. for three days; amphetamine, 5 mg/kg b.i.d. for three days. Note that whereas reserpine produces a significant increase in tyrosine hydroxylase, the antidepressant and stimulant drugs lead to a decrease.

latency, duration, and differential time course of various regions indicating axoplasmic flow (see following section) is highly suggestive. Figure 20 is a summary of the changes in specific activity of tyrosine hydroxylase determined at varying periods after the initiation of the tricyclic antidepressant—imipramine, the monoamine oxidase inhibitor—pargyline, the stimulants—amphetamine, and reserpine. Note that reserpine given in daily doses of 0.5 mg led to a significant *increase* in midbrain soluble tyrosine hydroxylase (50,000 x G supernatant) by eight days, whereas imipramine at three and eight days led to a significant *decrease* in this enzyme's specific activity (as did pargyline and amphetamine).

Figure 21 demonstrates the time course of the reserpine-induced effect and evidence that the concomitant malnutrition was not a significant variable. Recent studies following the specific activity of midbrain and caudate tyrosine hydroxylase after termination of reserpine treatment indicated that the raised levels were still maintained seven days later. This sort of duration is in marked contrast to the propranolol-induced increase lasting four hours (Fig. 12). We are in the process of following these animals still further. It is tempting to speculate about what this kind of time course might mean about the half-life of

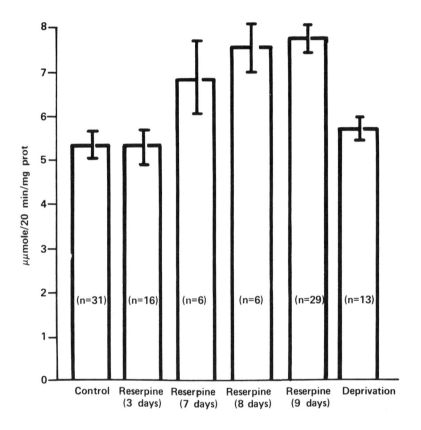

FIGURE 21. The effects of four schedules of reserpine treatment (see text) or food deprivation on the midbrain tyrosine hydroxylase activity of rats. Each bar represents the mean ± s.e.m. (brackets) of the indicated number of observations. Only eight to nine days of chronic reserpine treatment induced a significant increase in enzyme activity (p < 0.05, Mann-Whitney U test).

either the newly synthesized or increased amount of enzyme, but the prolonged and indeterminate nature of the effects of reserpine make this approach less than appropriate. Studies with inhibitors of protein and nucleic acid synthesis on similar reserpine-induced effects in the adrenal and sympathetic ganglion by Mueller, Thoenen, and Axelrod (1969) suggest that increases in the rate of macromolecular biosynthesis may be the mechanism for these kinds of changes. Studies reported here (Fig. 20) are the first in a brain adrenergic system which demonstrate a *decrease* as well as an increase in enzyme specific activity as an adaptation to drug treatment. Dairman and Udenfriend (1970) demonstrated this kind of change in the adrenal with DOPA loads. The fact that this level can move predictably in both directions is certainly consonant with its role as a potential regulatory mechanism. Other examples of these longer lasting changes induced by various agents and circumstances will be presented in a later section.

Evidence suggesting axoplasmic flow of neurotransmitter biosynthetic enzymes from cell body to nerve endings in the serotonergic and catecholamine systems. A considerable amount of work has been done relating nerve cell-body protein synthesis to the rate and method of its delivery to nerve endings (Barondes, 1967). In general, it has been concluded that most new protein in nerve endings (except for specialized nerve-ending mitochondrial protein) arrives via axoplasmic flow rather than local synthesis. Since we are beginning to see that chronic treatment with drugs or experimental conditions lead to adaptive and relatively long-lasting changes in levels of enzyme activity, we felt it would be significant to compare areas of high cell-body and nerve-ending density in one biogenic amine system to see if there was a systematic temporal relationship between changes in the cell-body and nerve-ending areas. For example, Figure 16 demonstrates that a single dose of PCPA leads to: (1) A quickly reversible decrease in substrate uptake in the nerve-ending, particulate septal enzyme (2) a more gradual and sustained decrease in midbrain, cell-body soluble enzyme as the irreversible inhibition via incorporation of PCPA into the enzyme protein occurs (Jequier, Levenberg, and Sjoerdsma, 1967) ; and (3) the late arrival of this defective enzyme to the septal nerve-ending area after several days.

Figure 22 represents the results of a similar study of a catecholamine system in which the reserpine-induced increase in tyrosine hydroxylase is first manifested in the midbrain, cell-body, soluble enzyme source and later the wave of apparently new enzyme arrives as the corpus striatum. Since the time course of the sequential changes is in days, it is likely that these enzymes travel with the rate usually assigned to "slow flow" (Barondes, 1967). Thus far we have no evidence that drugs or other conditions alter the rate of flow, although drug-induced "waves" of increased tyrosine hydroxylase activity have been reported (Axelrod, 1971; Mandell, 1970). These time-locked regional enzyme changes from areas dominated by cell bodies to those with high nerve-ending density can be considered additional evidence in favor of drug-induced changes in *amount* of the enzyme protein in that it is perhaps a bit easier to imagine transport of new enzyme protein than a wave of reversible activation. On the other hand, we have previously referred to studies of the axoplasmic flow of sulfated mucopolysaccharides (Elam et al., 1970) which we have shown capable of activation of tyrosine hydroxylase. This area of inquiry would certainly benefit from usable antityrosine hydroxylase antibody.

Synaptic function regulation via apparent alteration in central receptor sensitivity. The so-called denervation super-sensitivity phenomenon in which loss of presynaptic input into a synapse leads to a decrease in the threshold of the receptor to excitation by transmitter has been studied rather thoroughly in peripheral systems such as the phrenic nerve-diaphragmatic muscle preparation (Friedman, Jaffe, and Sharpless, 1969). The speculation has been that this alteration in sensitivity involves an increase in the amount or availability of receptor sites and recently Potter and his group (1972), using what appears to be a specific cholinergic receptor-binding toxin, have been able to demonstrate this kind of change directly. In the brain, little evidence for such changes has been found, although this explanation is frequently invoked to explain such drug-induced phenomena as tolerance and withdrawal.

Although we are aware of the complexity of the approach, Segal's innovative method for the chronic administration of putative neurotransmitters and drugs directly into the ventricular system in freely moving rats (1969) made it possible for us to examine whether an apparent post-synaptic sensitization occurred concomitant with some of our treatments. Figure 23 is a diagrammatic representation of the intraventricular infusion apparatus in which automated counting of rearings and crossings as indices of spontaneous motor behavior is carried out during administration of drugs or putative neurotransmitters. The first evidence that there was a demonstrable alteration in this index of receptor sensitivity appeared in the recent study of the interaction between thyroid hormone and intraventricular norepinephrine (Emlen, Segal, and Mandell, 1972). Rats treated with daily subcutaneous shots of 0.5 mg/kg Na-L-Thyroxine (Sigma) for ten days became hyperactive but, more interestingly, considerably potentiated to the behaviorally activating effect of intraventricularly administered

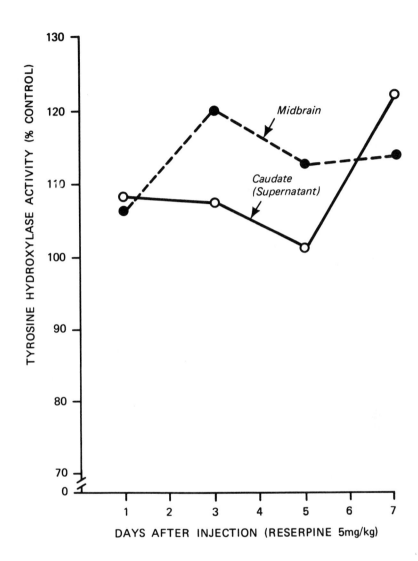

FIGURE 22.   The activity of midbrain and caudate tyrosine hydroxylase at various intervals following the administration of a single dose of reserpine, 5 mg/kg. There is an initial increase in both caudate and midbrain activity and then a later additional increase in the striate area following a return to control. The changes in the caudate can be seen as manifesting an initial enzymatic activation followed by the axoplasmic flow of new enzyme protein from the midbrain cell bodies to the striate nerve endings.

NE (given at a rate of 0.3 μg/min in a 1.0 μg/μl solution). Figure 23 demonstrates the mean increase in crossovers induced by NE with significantly higher values shown by the thyroid-treated animals. The effect of NE infusion can be seen to be potentially quite complex and capable of invoking a number of changes in addition to stimulation of the receptor. For example, it could be seen to be saturating the pre-synaptic storage pool or acting like tyramine and releasing endogenous biogenic amines. The thyroid-induced changes could be seen as influencing the degree of response of these presynaptic mechanisms. It was therefore of considerable interest to us to observe the same kind of sensitization to intraventricularly administered NE emerging gradually after the destruction of catecholamine nerve endings with 6-hydroxydopamine. In our studies, the first thing that happened was the

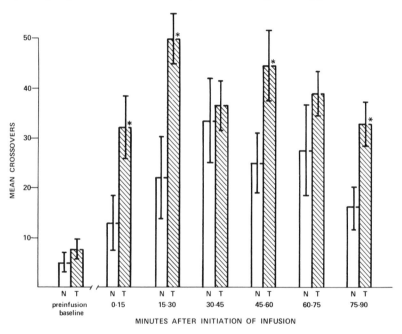

FIGURE 23. The effects of the infusion of 1.0 μg/μl NE (DL Arterenol-HCl) on the behavioral activity of nontreated (N) and thyroid-treated (T) rats. Crossovers were measured in 15-minute blocks for a 2.5-hour period. During the first hour, the animals received no infusion, and the last 15-minute block of this hour was used as the preinfusion base line. The data is presented as the mean number of crossovers ± SEN (brackets). Thyroid-treated (T) rats exhibited a potentiated response to NE infusion with respect to the onset, duration, and magnitude of the hyperactivity. * = Significantly different from controls: p < 0.01.

loss of striate and midbrain tyrosine hydroxylase activity followed by the appearance of supersensitivity to NE infusions identical to that seen following treatment with thyroid hormone. This change is consonant with the interpretation that a post-synaptic catecholamine receptor mechanism is responsible for the observed changes.

Evidence suggesting that cyclic-AMP may play a role in the regulation of neurotransmitter biosynthetic enzymes. Much recent work has highlighted the potential role of Sutherland, Robinson, and Butcher's "second messenger system" (1968) as an intracellular trigger for adaptive changes. We have started to explore the possibility that this mechanism may be an intervening variable in some of the changes in neurotransmitter biosynthetic enzymes. After establishing that

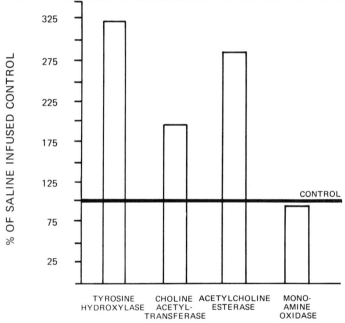

SPECIFIC ACTIVITY OF ENZYMES FROM CAUDATE NUCLEUS OF RAT FOLLOWING INTRAVENTRICULAR INFUSION OF CYCLIC-AMP

FIGURE 24. The specific activities of four caudate brain enzymes expressed as percent of saline control, 24 hours following the infusion of 23 micrograms of cyclic AMP over three hours. (See the text.)

dibutyrlcyclic-AMP did not alter enzyme activity *in vitro,* we infused 20 to 30 micrograms of this substance over a three-hour period (which produced a norepinephrine-like behavioral hyperactivity) and twenty-four hours later studied caudate tissue for the specific activity of tyrosine hydroxylase, choline acetyltransferase, acetylcholine esterase, and monamine oxidase. Figure 24 demonstrates an increase in those enzymes that we have seen to be alterable with long-term drug treatment. It may be that the report of Miyamoto, Kuo, and Greengard (1969) demonstrating the stimulation of brain protein kinase with cyclic-AMP can be the mechanism responsible for histone derepression and new brain enzyme synthesis. We are currently working on this sort of regulatory mechanism *in vivo* and with brain cell tissue culture to see if this approach may help us elucidate a sequential story of the process of brain enzyme adaptation.

Evidence suggesting that drug manipulations of one neurotransmitter system may produce changes in another system. Thus far, we have indicated that neurotransmitter enzyme regulatory changes of various kinds can occur when the system in which the enzyme functions is directly affected by drugs. The potential adaptational organization of these mechanisms will be discussed in a following section. In attempts to manipulate the specific activity of the acetylcholine synthesizing enzyme, choline acetyltransferase, we chose to use the optic lobes of two-week-old White Leghorn chicks. This animal has an immature blood brain barrier and therefore peripherally acting drugs can easily get into the brain. Initial attempts using acute and chronic treatment with cholinergic and anticholinergic drugs (atropine and neostigmine), though producing marked behavioral changes in the animals, did not alter the specific activity of optic-lobe choline acetyltransferase as did drugs more clearly acting in the biogenic amine systems such as reserpine, 6-hydroxydopamine, and methamphetamine. Figure 25 demonstrates increases in enzyme-specific activity after a single and several doses of methamphetamine, 10 mg/kg. Methamphetamine did not produce this activation *in vitro.* Acetoxycycloheximide given in doses which inhibit the incorporation of ^{14}C-leucine into TCA percipitable protein by 90 percent did not alter the specific activity of optic-lobe choline acetyltransferase over the six hours of the experiment. This is consonant with (but *not* proof of) an interpretation of an increase in new protein synthesis in response to methamphetamine. Monoamine oxidase determined at the same time manifested no changes.

Both the findings that a drug that may alter the neurotransmitter dynamics at one chemical kind of synapse may alter the presynaptic enzyme activity of another and our demonstration of the product-

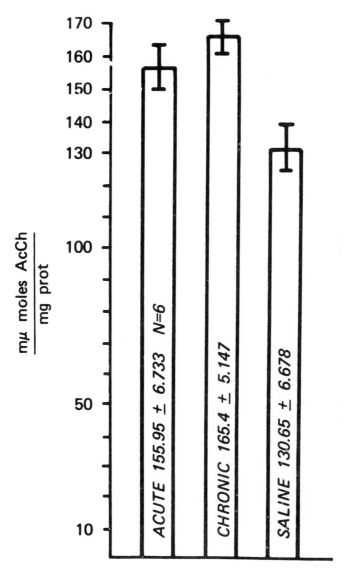

FIGURE 25. The specific activity of choline acetyltransferase in optic lobe following the acute and chronic administration of methadrine, 10 mg/kg., twice a day. The control group consisted of six chicks injected with 50 μl of saline, twice per day. Note the elevation of the specific activity following the acute and chronic administration of methadrine. The chronic methadrine values are significantly higher than the saline controls (p < .01).

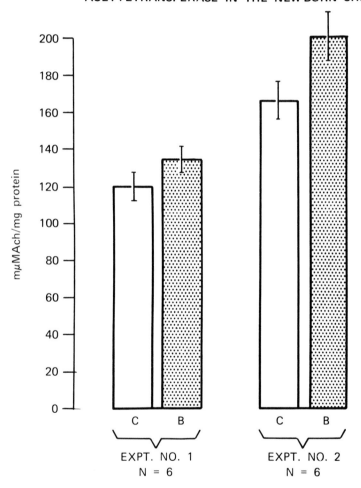

THE EFFECT OF BLINDING ON OPTIC LOBE CHOLINE
ACETYLTRANSFERASE IN THE NEW-BORN CHICK

FIGURE 26.   The effect of enucleation at birth on optic lobe choline acetyltransferase activity. C represents the lobes opposite the good eye; B represents the
lobes opposite the blinded eye. (See the text.)

feedback potential of one biogenic amine for another amine's enzyme system (Fig. 19) make model building within one system exclusively an untenable over-simplification. Although necessity requires at this time that we think and work within this simple context, it appears worthwhile to acknowledge the fact that puzzles of intersystem chemical regulation of enzymatic systems await us.

Studies relating environmental conditions and behavioral state to neurotransmitter biosynthetic enzyme activity

Studies suggesting that neurotransmitter enzymes may be environmentally sensitive compensatory mechanisms. One of our first approaches in an attempt to relate alterations in neurotransmitter biosynthetic enzyme activity to environmental triggers and behavioral states made use of the "disuse atrophy" hypothesis. That is, we guessed that if the brain systems worked like muscles, decreasing the demand for their activity might decrease their neurotransmitter enzyme complement. A propitious experimental locus for studying this question appeared to be the chick, in that the visual pathways from each eye are totally separate and it has prominent and clearly demarcated optic lobes. By blinding one eye, using the other side as a control, and studying choline acetyltransferase activity in both optic lobes we could approach this question. Figure 26 unexpectedly demonstrates that one week after blinding in two experiments, the "blinded" lobe was *higher* in specific activity in the enzyme that participates in the synthesis of acetylcholine than was the control. Although neither experimental change reached significance, the trend was observed in both experiments.

Approaching this issue using other enzyme systems, rats were maintained in quiet, dark quarters for up to sixteen days, sacrificed at various intervals, and the specific activities of midbrain and caudate tyrosine hydroxylase were compared with those values from animals kept in normal noise and light. The periods selected were four, eight, and sixteen days. Figure 27 indicates that rats kept in relative sensory isolation manifested significant increases in midbrain tyrosine hydroxylase by day 4 over control values. The striate area manifested this change with a delay (axoplasmic flow of new enzyme protein?). Thus it appears that environments via their arousal or sensory-input characteristics may lead to a change in neurotransmitter biosynhetic enzyme activity which may be opposite to what one might speculate to be the level of synaptic activity induced by the input.

Behavioral-genetic studies relating neurotransmitter biosynthetic enzyme levels to behavior. Five strains of inbred male rats were studied for spontaneous motor activity and the specific activity of midbrain

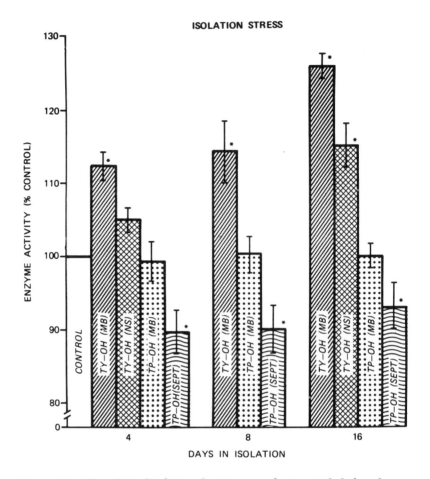

FIGURE 27. The effect of isolation of rats in a sound-attenuated, darkened room for 4, 8, and 16 days on midbrain (MB), neustriatal (NS), and septal (SEPT) tyrosine hydroxylase (TY-OH) and tryptophan hydroxylase (TP-OH) activities. Note that there is an immediate and progressive increase in midbrain tyrosine hydroxylase, a delayed increase in neostriatal tyrosine hydroxylase, and an immediate decrease in septal tryptophan hydroxylase.

and striatal tyrosine hydroxylase (see Materials and Methods section; Segal, Kuczenski, and Mandell, 1971). The results of these studies are summarized in Table 1 A, B, and C. The data reveal a wide range of interstrain spontaneous activity levels as measured by the number of crossovers (A). Rearings for the different strains were directly related to the order of crossovers. Similarly, a wide range of tyrosine hydroxylase levels in both striatum (B) and midbrain (C) was also apparent between the strains. In general, there was an inverse relationship between the mean spontaneous activity of each strain and the mean level of enzyme activity. BUF, which exhibited the highest number of crossovers, had the lowest level of enzyme. Conversely, F344, the least active of six strains, was found to have the highest level of enzyme activity. The strains intermediate in activity showed a similar though less dramatic inverse relationship with enzyme level. The correlation between the behavior and striatal tyrosine hydroxylase was -0.94 which was significant at $p < .02$; the correlation between the behavior and the midbrain enzyme was -0.83 ($p < .08$). Again, an inverse relationship was observed between a major catecholamine biosynthetic enzyme-specific activity and what can be assumed to be a behavioral reflection of increased catecholamine synaptic activity. This kind of relationship again appears to be a compensatory one in which it is likely that increases in synaptic activity may be associated with a feedback mediated decrease in biosynthetic enzyme activity. Both the measurements of behavior and enzyme-specific activity appear to be partial correlations involving a third and as yet undetermined system. The similarity of the inverse relationship between neurotransmitter enzyme activity and spontaneous motor activity that is maintained through both environmental and genetically altered circumstances makes a rather strong case for this relationship (although some drug-induced exceptions will be seen).

Studies varying thyroid hormone relating neurotransmitter enzyme activity to behavior. In addition to thyroid hormone administration leading to apparent sensitization of the catecholamine receptor (Figure 24), the thyroid treatment increased behavioral activity but was *not* associated with a decrease in midbrain tyrosine hydroxylase (Fig. 28). However, the inverse state, a decrease in circulating thyroid hormone three weeks following thyroidectomy, led to an increase in midbrain tyrosine hydroxylase associated with a decrease in spontaneous motor activity. This latter reciprocal relationship is consonant with the findings in previously reported environmental and genetic studies. This study is also consonant with Prange's findings that thyroidectomy increases the rate of tyrosine to catecholamine turnover (Prange, Meek, and Lipton, 1970). It is of interest to note that alterations in receptor sensitivity and presynaptic neurotransmitter enzyme activity alterations may not necessarily be linked as concomitant adaptations in that they appear to be separable in the thyroid-treated animals.

Studies using psychotropic drugs relating neurotransmitter biosynthetic enzyme activity to behavior. In a previous section, we re-

Table 1. STRAIN DIFFERENCES IN LEVELS OF TYROSINE HYDROXYLASE AND SPONTANEOUS ACTIVITY

		A. SPONTANEOUS ACTIVITY				
Strain		BUF	SD	LEW	ACI	BN	F344
		137*	121	114	92	59	46
		± 29	± 27	± 24	± 29	± 15	± 4
		(n = 7)	(n = 7)	(n = 7)	(n = 7)	(n = 7)	(n = 7)
BUF	137 ± 29	—	—	—	—	—	—
SD	121 ± 27	N.S.	—	—	—	—	—
LEW	114 ± 24	N.S.	N.S.	—	—	—	—
ACI	92 ± 29	N.S.	N.S.	N.S.	—	—	—
BN	59 ± 15	< .01	< .05	< .05	N.S.	—	—
F344	46 ± 4	< .01	< .001	< .001	< .05	N.S.	—

B. STRIATAL TYROSINE HYDROXYLASE

		BUF	SD	LEW	BN	ACI	F344
Strain		12,485†	13,919	14,867	15,079	15,572	17,944
		± 556	± 406	± 265	± 982	± 902	± 418
		(n = 10)	(n = 15)	(n = 15)	(n = 5)	(n = 10)	(n = 15)
BUF	12,485 ± 556	—	—	—	—	—	—
SD	13,919 ± 406	< .01	—	—	—	—	—
LEW	14,867 ± 265	< .001	< .01	—	—	—	—
BN	15,079 ± 982	N.S.	N.S.	N.S.	—	—	—
ACI	15,572 ± 902	< .01	< .05	N.S.	N.S.	—	—
F344	17,944 ± 418	< .001	< .001	< .001	< .05	< .002	—

C. MIDBRAIN TYROSINE HYDROXYLASE

		BUF	ACI	SD	LEW	BN	F344
Strain		1,596†	1,707	1,830	1,831	2,432	2,613
		± 139	± 75	± 122	± 120	± 100	± 66
		(n = 5)	(n = 10)	(n = 15)	(n = 10)	(n = 5)	(n = 5)
BUF	1,596 ± 139	—	—	—	—	—	—
ACI	1,707 ± 75	N.S.	—	—	—	—	—
SD	1,830 ± 122	N.S.	N.S.	—	—	—	—
LEW	1,831 ± 120	N.S.	N.S.	N.S.	—	—	—
BN	2,432 ± 100	< .004	< .001	< .01	< .01	—	—
F344	2,613 ± 66	< .004	< .001	< .01	< .001	N.S.	—

Levels of significance were calculated using the Mann-Whitney U test (Siegel, 1957).
* Mean number of cross-overs ± S.E.M.
† Mean net cpm ^3H_2O released/20 min/mg protein ± S.E.M.

The effects of thyroid treatment and thyroidectomy on the specific activity of midbrain tyrosine hydroxylase (expressed as percent of control). Thyroid-treated animals received 0.5 mg/kg Na-L-thyroxine for ten days, and showed no change in enzyme activity. Thyroidectomized animals were sacrificed three viewed some of our results using behaviorally depressing, activating, and antidepressant drugs with reference to midbrain tyrosine hydroxylase activity (Fig. 20). We characterized these kinds of drug-induced changes as requiring lower doses, maintenance of tissue levels, longer latency to full effect, and rather considerable amounts of time (days) to return to control levels. A dramatic and apparently paradoxical experiment demonstrating a *direct* relationship between the activity of tyrosine hydroxylase and motor behavior (Segal et al., 1971) consisted of the chronic treatment of rats with reserpine, 0.5 mg/day for nine days, monitoring the spontaneous motor activity as outlined previously. Figure 29 summarizes the effects of two schedules of reserpine treatment (three days versus nine days) or food deprivation on motor movement monitored one hour per day for nine days. Note that marked behavioral hyperactivity ensued after behavioral depression with the transition between five and seven days. A comparison with Figure 21 demonstrates that the increase in motor activity almost precisely fits in time the *increases* in tyrosine hydroxylase activity. At first glance, these results appear not to follow the inverse relationship between this enzyme's activity of spontaneous motor behavior noted previously in the environmental genetic and hormonal studies. This is particularly difficult to interpret in that treatment with other drugs, such as the antidepressant imipramine as well as the MAOI's and amphetamines, chronically led to a decrease in tyrosine hydroxylase and increases in spontaneous motor activity (consonant with the reciprocal relationship described in other studies).

In the case of chronic reserpine treatment, we seem to be dealing with the initiation of a secondary, extraordinary adaptive attempt following the failure of the first-line adaptive attempt by the neurotransmitter biosynthetic enzymatic machinery. Reserpine, in a higher acute dose (5 mg/kg) produces a marked behavioral depression associated with an *increase* in tyrosine hydroxylase activity with a considerably faster rate of increment (Fig. 13). This same kind of fast enzymatic *activation* associated with decreased motor activity was seen following the acute administration of propranolol (Fig. 30). We have speculated that whereas the propranolol-induced enzymatic activation could be as-

weeks after thyroidectomy and showed a significant increase in enzyme activity ($p < 0.01$). Brackets represent the standard error of the mean. The effects of thyroid treatment on free-field behavioral activity. Activity during a 75 minute period was measured as crossovers from one quadrant to another. Thyroid-treated rats showed a significant increase in activity ($* = p < 0.02$), and thyroidectomized rats showed a decrease in activity, although not significant ($** = 0.05 < p < 0.1$). Brackets represent the standard error of the mean.

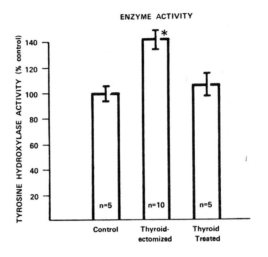

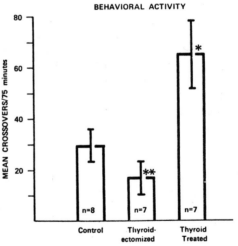

FIGURE 28. *See Mandell, et al, p. 53.* (Top) The effects of thyroid treatment and thyroidectomy on the specific activity of midbrain tyrosine hydroxylase (expressed as % control). Thyroid treated animals received 0.5mg/kg Na-L-thyroxine for 10 days, and showed no change in enzyme activity. Thyroidectomized animals were sacrificed 3 weeks after thyroidectomy and showed a significant increase in enzyme activity ($p < 0.01$). Brackets represent the standard error of the mean. (Bottom) The effects of thyroid treatment on free field behavioral activity. Activity during a 75 min. period was measured as cross-overs from one quadrant to another. Thyroid treated rats showed a significant increase in activity (* = $p < 0.02$), and thyroidectomized rats showed a decrease in activity, although not significant (** = $0.05 < p < 0.1$). Brackets represent standard error of the mean.

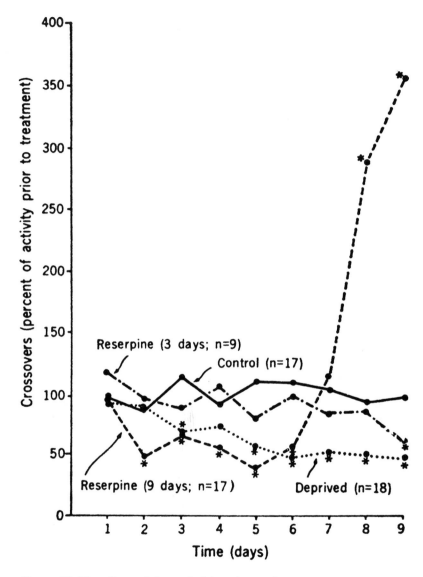

FIGURE 29. The effects of four schedules of reserpine treatment (see text) or food deprivation on the midbrain tyrosine hydroxylase activity of rats. Each bar represents the mean ± s.e.m. (brackets) of the indicated number of observations. Only eight to nine days of chronic reserpine treatment induced a significant increase in enzyme activity (p < 0.05, Mann-Whitney U test).

sociated with increments in intraneuronal NE and therefore regulation by product-feedback, the acute reserpine-induced activation of tyrosine hydroxylase associated with a reserpine-promoted loss of intraneuronal NE would lead to an activated enzymatic state that was *not* associated with either increments in or return to normal of intraneuronal NE. This prolonged failure of compensation (i.e., return to normal levels of intraneuronal NE) due to the reserpine-destroyed catecholamine storage processes or loss of vesicular DBH (Viveros et al., 1969) might be the trigger for another order of adaptive response (new enzyme protein synthesis) which is less finely tuned and could lead to an overshoot, "pathological" activation and hyperactivity.

This difference between physiological adaptational relationships in the catecholamine biosynthetic systems (such as the environmental, genetic, thyroid-induced, and antidepressant drug-induced reciprocal relationships) and the chronically administered reserpine-induced phenomenon might have been predicted from the course of the adaptive

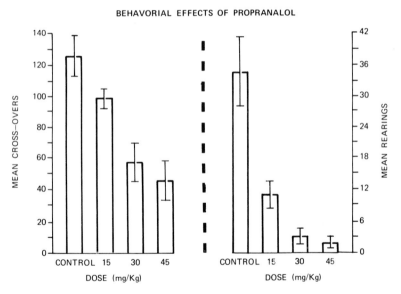

BEHAVORIAL EFFECTS OF PROPRANALOL

FIGURE 30. The effects of increasing doses of propranolol on two parameters of behavioral activity measured during the one-hour time interval immediately following parenteral drug administration. A dose-dependent behavioral depression which paralleled the increase in striatal tyrosine hydroxylase activity was observed for both crossovers and rearings. With the exception of crossovers at the 15 mg/kg dose, all values were significantly lower than control (p < .002), n = 9 for each group.

resynthesis of depleted catecholamines in the adrenal produced by electrical stimulation and low doses of insulin compared to the results following massive depletion for days followed by "overshoot" triggered by massive stimulation by high doses of insulin, nicotine, or acetylcholine discussed in the early pages of this paper. It may be that especially in those areas with particulate or bound tyrosine hydroxylase which resists reserpine depletion the return to more physiological mechanisms of regulation and to the reciprocal relationship between tyrosine hydroxylase activity and behavior await the return of the storage and binding capacity of nerve endings for catecholamines or the vesicular-bound enzyme, dopamine-β-hydroxylase. This hypothesis, currently being explored, is particularly important in that it touches on the issue of the signal of most significance in triggering these adaptive responses—the information the receptor receives versus the chemical state of the intraneuronal presynaptic apparatus sending the information. Dairman and Udenfriend's (1970) work demonstrating a decrease in adrenal (but not brain) tyrosine hydroxylase following DOPA cannot be explained simply by a neurally mediated feedback mechanism in that decentralization by itself fails to decrease adrenal tyrosine hydroxylase activity.

Although we have argued here for the critical role of intraneuronal catecholamine concentration to explain signals leading to adaptive changes in neurotransmitter biosynthetic enzymes, one could argue that the post-synaptic receptor state and accompanying corrective feedback to the presynaptic neuron is the more appropriate model. Certainly the trans-synaptic enzymatic adaptations in peripheral adrenergic systems as reviewed by Thoenen and others (1961) appear to be best understood this way. This later point of view would lead to seeing the activity of the neurotransmitter enzyme as a mirror image of the functional level of synaptic neurotransmitter such that genetic, environmental, hormonal, and drug-induced increases in synaptically active neurotransmitter or receptor sensitivity would be followed by a feedback signaled alteration in the amount or activity of the neurotransmitter enzyme. In this case, the *direct* relationship between motor activity and midbrain tyrosine hydroxylase activity induced by chronic reserpine treatment would indicate that the increase in tyrosine hydroxylase and biosynthesis of norepinephrine (Weiner, 1970) might not increase synaptically active transmitter since the transmitter is unprotected from cytoplasmic monoamine oxidase by intact vesicular storage mechanisms; or dopamine is not converted to norepinephrine due to reserpine-induced loss of vesicular DBH (Viveros et al., 1969). We have recently demonstrated that DA is considerably less behavior-

TREATMENT	RAT MIDBRAIN TYROSINE HYDROXYLASE ACTIVITY	MOOD IN MAN
RESERPINE	↑	↓
PROPRANOLOL	↑	↓
HYPOTHYROIDISM	↑	↓
IMIPRAMINE	↓	↑
PARGYLINE	↓	↑
AMPHETAMINE	↓	↑

FIGURE 31. A summary of the relationship between drug effects on midbrain (propranolol on striate) tyrosine hydroxylase activity and their effect on mood in man. Note that there appears to be an inverse relationship. (See the text.)

ally activating than is NE (Geyer, 1971) when administered intraventricularly.

Both the model that emphasizes presynaptic intraneuronal catecholamine content and the one that focuses on post-synaptic receptor activity can explain such effects as that seen with imipramine. Imipramine potentiates synaptic catecholamine as well as impeding the spontaneous release of presynaptic catecholamines. We would predict that both or either of these effects would lead to a decrease in amount or activity of presynaptic neurotransmitter biosynthetic enzyme. Effects such as that seen with sensory isolation appear to be less tortuously explained by receptor activity as the significant trigger.

A New Neuropharmacological Theory of Depression

In the development of neurochemical models for psychiatric disorder, the most common research approach is the use of what Schildkraut has called the pharmacological bridge (1969). After it became quite clear that studies of relevant-sounding metabolites in body fluids are too "noisy" to reflect the metabolism of the brain (Mandell, 1969), we have all turned to this "Bridge" in an attempt to link neurochemical models derived in animals to behavior in man. In simple terms, this approach assumes that the psychotropic drug-induced changes in neurochemical parameters studied in animal brains underlie the behavioral alterations that these drugs induce in man. Up to this time, the most commonly used chemical parameters have been drug-induced alterations in the turnover or metabolite patterns of intraventricularly administered, isotopic biogenic amines (Schildkraut, 1969). Conclusions from these studies have tended to center around the generalization that increases in synaptically active neurotransmitter (catecholamines) are associated with elevated mood and the reverse.

Our work relating neurotransmitter biosynthetic enzyme activity and spontaneous motor-activity to various pharmacological and non-pharmacological treatments, has led us to what appears to be another systematic group of chemical and behavioral observations which may not necessarily be at odds with the previous theory in all ways, but appear to add predictions and dimensions that are not totally accounted for in the previous theory. The increasing emphasis of recent research on neurotransmitter biosynthesis as regulating functional neurotransmitter levels and the emerging myriad of exquisitely responsive regulatory mechanisms affecting these processes bespeaks of the potential importance of this approach. In addition, the latency of action of particularly the antidepressant drugs, though they appear to affect synaptic mechanisms immediately, suggests that perhaps adaptational mechanisms (of the sort we have been studying) requiring a longer latency may be the relevant chemical measures. Such drugs as imipramine and even the monoamine oxidase inhibitors take days to weeks to work, although measurable alterations in synaptic mechanisms occur within hours.

In its simplest form, our model states that midbrain tyrosine hydroxylase activity levels are elevated in those conditions associated with clinical depression. The two most valid appearing, inducible depressive syndromes in man (associated with sleep loss, vasovegetative disturbance, ruminative self-derogation, etc.) follow upon the chronic administration of reserpine or spontaneously appearing (or iatrogenic) hypothyroidism (Bein, 1970; Prange, Meek, and Lipton, 1970). Both of these treatments elevate midbrain and striate tyrosine hydroxylase. Propranolol, which has been reported to produce severe depression (Sullivan et al., 1971) in man, also increases tyrosine hydroxylase. The antidepressant, imipramine, and to a lesser extent the monoamine oxidase inhibitors, pargyline and amphetamine, decrease midbrain tyrosine hydroxylase activity. The amount that they decrease this activity appears nicely correlated with their relative efficacy as antidepressants in severe cases (Fig. 31). Thus, drug treatments associated with induction of clinical depression are associated with an increase in this catecholamine biosynthetic enzyme. One can even get more poetic and assume that the sensory isolation experiment indicates that a complex and stimulating environment can be antidepressant (decrease midbrain tyrosine hydroxylase). Our preliminary studies suggesting an EST induced *decrease* in this midbrain enzyme also fit this suggested relationship. Our genetic studies suggest that if such a parameter as midbrain tyrosine hydroxylase activity level is relevant (and we are suggesting that it is), we have clear-cut evidence that it can be hereditary and related to a spontaneous motor-behavior pattern. In addition, in a recent series of studies we have shown that both drug-induced increases and decreases in this enzyme occur to about the same extent on a percentage basis in the various rat strains. This might indicate that if a critical level of a neurotransmitter enzyme is required to produce clinical depression, we may have experimental evidence for genetically based proneness. Since our most recent work indicates that on an inter-individual basis tyrosine hydroxylase varies *with* choline acetyltransferase, *inversely* with particulate tryptophan hydroxylase, and independently of catechol-O-methyltransferase and monoamine oxidase, it appears that these genetically determined neurochemical relationships may be relatively specific.

From a clinical point of view, it is tempting to look back over the syndromes of severe depression in light of this new neurochemical hypothesis. The midbrain catecholamine system has been viewed neurophysiologically as an "arousal system" for a number of years (Lindsley, 1960). In addition, we have shown that intraventricularly administered norepinephrine produces behavioral arousal on a dose-related basis (Segal and Mandell, 1970). With this sort of set, one can re-examine depressive symptomatology, highlighting some of its earliest and most pathogmonic features: insomnia, driven and obsessionally torturing thoughts, loss of appetite and weight loss, and even though immobile, these patients' minds are never still. Imipramine while it aids significantly a large percentage of these cases if they are carefully selected (Kuhn, 1970) shows some of its early effects by aiding sleep. It is worthwhile remembering that as far back as 1964, Schildkraut and others related peripheral signs of a *decrease* in catecholamine biosynthesis to the clinical response of patients to the tricyclic antidepressant. Using the reserpine and hypothyroid model, could these patients be suffering from a "pathological activation syndrome"? Could the remarkable hyperactivity of the chronically reserpinized rats represent the ceaseless drives of the brain of a depressive? Could the antidepressants work by decreasing the pathologically elevated tyrosine hydroxylase and therefore aiding the depressed patient? Could severe stresses in genetically prone people lead to at first a massively depleted intraneuronal catecholamine pool followed by a reserpine-like delayed increase in tyrosine hydroxylase activity and a "pathological activation syndrome"? Could decreases in receptor sensitivity (hypothyroid, receptor blocker-thorazine) lead to a compensatory increase beyond physiological limits of this potentially pathological enzymatic adaptation?

Some of the implications of this approach are currently being explored. For example, it may be that midbrain tyrosine hydroxylase levels can serve as a dependent variable in searches for new and effective antidepressants. Davis (1971) has recently told us about a new tricyclic antidepressant that does not block catecholamine uptake by synaptosomes like imipramine though it is an effective antidepressant. Soon we are going to be examining this drug's effect on neurotransmitter biosynthetic enzymes. In addition, we are engaged in studies of spinal fluid and white cell neurotransmitter biosynthetic enzyme activity in various psychiatric syndromes in the hope that if genetic tendencies for affect disorder are expressed in neurotransmitter biosynthetic enzyme complement, these tendencies can be seen in peripheral tissue and correlated to psychopathological syndromes.

Conclusion

Studies have been presented elucidating a number of potential regulatory mechanisms influencing the brain's neurotransmitter biosynthetic enzymes and receptors, including physical-state changes, allosteric alterations of the active site, product-feedback regulation, regulation of substrate availability, substrate activation of enzyme, apparent new enzyme protein biosynthesis, axoplasmic flow of new protein, alterations in receptor sensitivity, compensatory changes in one neurotransmitter system induced by changes in another, and the potential role of cyclic-AMP in the intracellular transduction of enzyme-altering information. These changes have been related to behavioral state by varying environmental, genetic, hormonal, and pharmacological variables and have elucidated an apparent systematic relationship. It has been suggested that either presynaptic, intraneuronal neurotransmitter levels or the state of the receptor (or both) are the significant signals for these short- and long-term enzymatic adaptations. Finally, it has been suggested that an enzymatic alteration (increases in midbrain tyrosine hydroxylase activity) may be a significant feature in clinical depression and explanatory of the pathogenic and therapeutic effects of drugs associated with causing and curing clinical depression. The implications of this model are being explored both for screening new drugs and in genetic studies of psychopathological groups in man.

ACKNOWLEDGMENT: This work was supported by NIMH Grants #MH-18065 and MH-14360 and the Benjamin Graham Foundation Research Fund.

Literature Cited

Axelrod, J. 1971. Noradrenaline: Fat and control of its biosynthesis. Science, 173: 598-606.

Barondes, S. H. 1967. Axoplasmic transport. NRP Bull., 5: 307-419.

Bien, H. J. 1970. Biological research in the pharmaceutical industry with reserpine. In Discoveries in Biological Psychiatry, F. J. Ayd and B. Blackwell, eds. Philadelphia: J. B. Lippincott.

Besson, M. J., A Cheramy, and J. Glowinski. 1971. Effects of some psychotropic drugs on dopamine synthesis in the rat striatum. J. Pharm. Exp. Therap., 177: 196-205.

Butterworth, K. R., and M. Mann. 1957. The release of adrenaline and noradrenaline from the adrenal gland of the cat by acetylcholine. Brit. J. Pharmacol., 12: 422-426.

Bygdeman, S., and U. S. von Euler. 1958. Resynthesis of catechol hormones in the cat's adrenal medulla. Acta Physiol. Scan., 44: 375-383.

Christensen, J. D., W. Dairman, and S. Udenfriend. 1970. Preparation and properties of a homogeneous aromatic L-amino acid decarboxylase from hog kidney. Arch. Biochem. and Biophys., 141: 356-357.

Cotman, C. W., and D. A. Flansburg. 1970. An analytical micro-method for electron microscope study of the composition and sedimentation properties of subcellular fractions. Brain Res., 22: 152-156.

Dairman, W., and S. Udenfriend. 1970. Increased conversion of tyrosine to catecholamines in the intact rat following elevation of tissue tyrosine hydroxylase levels by administered phenoxybenzamine. Molecular Pharmac., 6: 360-365.

Davis, J. 1971. Personal communication.

Elam, J. S., J. M. Goldberg, N. S. Radin, and B. W. Agranoff. 1970. Rapid axonal transport of sulfated mucopolysaccharide proteins. Science, 170: 458-459.

Emlen, W., D. S. Segal, and A. J. Mandell. 1971. Effects of thyroid state on pre- and post-synaptic central noradrenergic mechanisms. Science, in press.

Friedman, M. J., J. H. Jaffe, and S. H. Sharpless. 1969. Central nervous system supersensitivity to pilocarpine after withdrawal of chronically administered scopolamine. J. Pharmacol. Exp. Ther., 167: 45-55.

Gal, E. M., M. Morgan, and F. D. Marshall, Jr. 1965. Studies on the metabolism of 5-hydroxytryptamine (serotonin). IV. The effect of various drugs on the in vivo hydroxylation of tryptophan by the brain tissue. Life Sci., 4: 1765-1772.

Geyer, M. 1971. Ph.D. thesis in preparation, University of California at San Diego.

Grahame-Smith, D. G. 1964. Tryptophan hydroxylase in brain. Biochem. and Biophys. Res. Comm., 16: 586-592.

Grahame-Smith, D. G. 1968. Discussion of tryptophan hydroxylation in mammalian systems. Adv. Pharm., 64: 37-42.

Grahame-Smith, D. G. 1971. Studies in vivo on the relationship between brain tryptophan, brain 5-HT synthesis and hyperactivity in rats treated with a monoamine oxidase inhibitor and L-tryptophan. J. Neurochem., 18: 1053-1066.

Gray, E. G., and V. P. Whittaker. 1962. The isolation of nerve endings from brain: An Electromicroscope study of cell fragments derived by homogenization and centrifugation of cell fragments. J. of Anatomy, 96: 79-88.

Grungard, A., and P. Fergelson. 1961. The activation and induction of rat liver trypophan pyrolase *in vivo* by its substrate. J. Biol. Chem., *236:* 158.

Hillarp, N. A. 1960. Some problems concerning the storage of catecholamines in the adrenal medulla. In *Adrenergic Mechanisms,* pp. 481-486. Boston: Little Brown.

Hillarp, N., K. Fuxe, and A. Dahlstrom. 1966. Central monoamine neurons. In *Mechanisms of Release of Biogenic Amines,* U. S. von Euler, S. Rosell, and B. Uvnas, eds., pp. 31-58. New York: Pergamon Press.

Hokfelt, B., and J. McLean. 1950. The adrenaline and noradrenaline content of the suprarenal glands of the rabbit under normal conditions and after various forms of stimulation. Acta Pysiol. Scan., *21:* 258-270.

Holland, W. C., and H. J. Schumann. 1956. Formation of catecholamines during splanchnic stimulation of the adrenal gland of the cat. Brit. J. Pharmacol., *6:* 289-293.

Ichiyama, A., S. Nakamura, Y. Nishizuka, and O. Hyaish. 1970. Enzymic studies on the biosynthesis of serotonin in mammalian brain. J. of Biol. Chem., *245:* 1699-1709.

Jequier, E., W. Levenburg, and A. Sjoerdmsa. 1967. Tryptophan hydroxylase inhibition: The mechanism by which p-chlorophenalamine depletes rat brain serotonin. Mol. Pharmacol., *3:* 274-278.

Kopin, I. J., G. R. Breese, K. R. Krauss, and V. K. Weise. 1968. Selective release of newly synthesized norepinephrine from the cat spleen during sympathetic nerve stimulation. J. Pharmacol. Exp. Ther., *161:* 271-278.

Kosland, D. E., Jr. 1970. The molecular basis for enzyme regulation. In *The Enzymes: Structure and Control,* Vol. 1., P. D. Boyer, ed. New York: Academic Press.

Kuczenski, R. T., and A. J. Mandell. 1971. Allosteric activation of hypothalmic tyrosine hydroxylase by ions and sulfated mucopolysaccharides. J. Neurochem., in press.

Kuhn, R. The imipramine story. In *Discoveries in Biological Psychiatry.* 1970. F. J. Ayd and B. Blackwell, eds. Philadelphia: J. B. Lippincott.

Levitt, M., S. Spector, A. Sjoerdsma, and S. Udenfriend. 1965. Elucidation of the rate-limiting step in norepinephrine biosynthesis in the perfused guinea pig heart. J. of Pharmac. and Exper. Therap., *148:* 1-8.

Lindsley, D. B. 1960. Attention, consciousness, sleep and wakefulness. In *Handbook of Physiology,* Section I: Neurophysiology, J. Field, H. W. Magoun, and V. E. Hall, eds. American Physiological Society.

Lowry, O. H., N. J. Rosebrough, A. L. Fan, and R. J. Randell. 1951. Protein measurements with the folinphenol reagent. J. Biol. Chem., *193:* 265-273.

Mandell, A. J. 1969. Hormonal and metabolic correlates of behavioral states in man. In *Psychochemical Research in Man,* A. J. Mandell and M. P. Mandell, eds. New York: Academic Press.

Mandell, A. J. 1970. Drug-induced alterations in brain biosynthetic enzyme activity—a model for adaptation for the environment by the central nervous system. In *Biochemistry of Brain and Behavior,* E. Datta, ed. New York: Plenum Press.

Mandell, A. J., and M. Morgan. 1970. Amphetamine-induced increase in tyrosine hydroxylase activity. Nature, *227:* 75-76.

Mueller, R. A., H. Thoenen, and J. Axelrod. 1969. Inhibition of trans-synaptically increased tyrosine hydroxylase activity by cycloheximide and actinomycin D. Mol. Pharmacol., 5: 463-469.

Musacchio, J. M., L. Julou, S. S. Kety, and J. Glowinski. 1969. Increase in rat brain tyrosine hydroxylase activity produced by electroconvulsive shock. Pro. Nat. Acad. Sci., 63: 1117-1119.

Myamoto, E., J. T. Kiso, and P. Greengard. 1969. Adenosine 3, 5-monophate-dependent protein kinase from brain. Science, 165: 63-65.

Nagatsu, T., M. Levitt, and S. Udenfriend. 1964. Tyrosine hydroxylase: The initial step in nonrepinephrine biosynthesis. J. Biol. Chem., 239: 2910-2917.

Perez-Cruet, J., A. Tageiamonte, P. Tageiamonte, and G. L. Gessa. 1971. Stimulation of serotonin synthesis by lithium. JPET, 178: 325-330.

Porcellati, G., and F. di Jeso. 1971. Membrane-Bound Enzymes. New York: Plenum Press.

Potter, L., and P. B. Molinoff. 1972. Isolation of cholinergic receptor protein. In Perspectives in Neuropharmacology, S. H. Snyder, ed. Oxford: Oxford University Press, in press.

Prange, A. J., J. L. Meek, and M. A. Lipton. 1970. Catecholamines: Diminished rate of synthesis in rat brain and heart after thyroxine pretreatment. Life Sci., 9: 901.

Schier, B. K., and L. Shuster. 1967. A simplified radiochemical assay for choline acetyltransferase. J. Neurochem., 14: 977-984.

Schildkraut, J. J. 1969. Rationale of some approaches used in biochemical studies of the affective disorders: The pharmacological bridge. In Psychochemical Research in Man, A. J. Mandell and M. P. Mandell, eds. New York: Academic Press.

Schildkraut, J. J., G. L. Lerman, R. Hammond, and D. G. Friend. 1964. Execretion of 3-methoxy-4-hydroxymandelic acid (VMA) in depressed patients treated with antidepressant drugs. J. Psychiat. Res., 2: 257-266.

Sedvall, G. C., V. K. Weise, and I. J. Kopin. 1968. The rate of norepinephrine synthesis measured in vivo during short intervals; influence of adrenergic nerve impulse activity. J. of Pharmac. and Exper. Therap., 159: 274-282.

Segal, D. S. 1969. Catecholamine and Behavior. Ph.D. thesis, University of California at Irvine.

Segal, D. S., and A. J. Mandell. 1970. Behavioral activation of rats during intraventricular infusion of norepinephrine. Proc. of Nat. Acad. Sci., 66: 289-293.

Segal, D. S., R. T. Kuczenski, and A. J. Mandell. 1971. Strain differences in behavior and brain tyrosine hydroxlase activity. Behavioral Biol., in press.

Segal, D. S., J. L. Sullivan, R. T. Kuczenski, and A. J. Mandell. 1971. Effects of long-term reserpine treatment on brain tyrosine hydroxylase and behavioral activity. Science, 173: 847-848.

Siegel, S. 1956. Nonparametric Statistics, pp. 116-127. New York: McGraw Hill.

Sullivan, J. L., D. S. Segal, R. T. Kuczenski, and A.J. Mandell. 1971. Propranolol induced rapid activation of rat striatal tyrosine hydroxylase concomitant with behavioral depression. Biol. Psychiatry, in press.

Sutherland, E. W., G. A. Robinson, and R. W. Butcher. 1968. Some aspects of the biological role of adenosine-3, 5-monophosphate (cyclic-AMP). Circulation, 3: 279-306.

Thoa, N. B., D. G. Johnson, I. J. Kopin, and N. Weiner. 1971. Acceleration of catecholamine formation in the guinea-pig vas deferens after hypogastric nerve stimulation: Roles of tyrosine hydroxylase and new protein synthesis. J. of Pharmac. and Exper. Therap., 178: 442-449.

Thoenen, H. 1970. Induction of tyrosine hydroxylase in peripheral and central adrenergic neurons by cold-exposure of rats. Proc. Nat. Acad. Sci., *63:* 1117-1119.

Thoenen, H., R. Kettler, W. Burkard, and A. Saner. 1971. Neurally mediated control of enzymes involved in the synthesis of norepinephrine: Are they regulated as an operational unit? Naunyn-Schmiedebergs Arch. Pharmak., *270:* 146-160.

Udenfriend, S., J. R. Cooper, C. T. Clark, and J. E. Baer. 1953. Rate of turnover of epinephrine in the adrenal medulla. Science, *117:* 663-665.

Viveros, O. H., L. Argueros, R. J. Connet, and N. Kirshner. 1969. Mechanism of release from the adrenal medulla IV. The fate of storage vescicles following insulin and reserpine administration. Mol. Pharm., *5:* 69-79.

von Euler, U. S., and F. Lishajko. 1966. Inhibitory action of adrenergic blocking agents on catecholamine release and uptake in isolated nerve granules. Acta Physiol. Scan., *68:* 257.

Vos, J., K. Kuriyama, and E. Roberts. 1968. Electrophoretic mobilities of brain subcellular particles and binding of Q-aminobytyric acid, acetylcholine, NE, and 5-HT. Brain Res., *9:* 224.

Weiser, N. 1970. Regulation of norepinephrine brosynthesis. *Ann. Rev. Pharm.* Stanford: Stanford University Press.

West, G. B. 1951. Insulin and the suprarenal gland of the rabbit. Brit. J. Pharmacol., *6:* 289-293.

Wurzburger,R. J., and J. M. Musacchio. 1971. Subcellular distribution and aggregation of bovine adrenal tyrosine hydroxylase. JPET, *177:* 155-167.

Neural Activity and Behavioral Plasticity in the Crayfish Claw

G. L. Gerstein, R. H. Schuster, and T. J. Wiens
Departments of Physiology and of Biophysics
School of Medicine
University of Pennsylvania
Philadelphia, Pennsylvania

The subtleties of the nervous system have occupied investigators for well over a century, and a tremendous amount of anatomical, electrophysiological, and biochemical information has been amassed. In spite of all this work, a number of the most striking properties of the nervous system remain to be understood completely. One set of such properties determines the nature of memory consolidation, storage, and recall; this is reviewed in one of the other papers of this volume (Agranoff). Another property, which will be the concern here, is the ability of the nervous system to reorganize or alter itself as the result of experience. Such alterations have been given the name "plasticity."

Plasticity has been recognized in a wide variety of animals both behaviorally and electrophysiologically, with varying degrees of permanence and varying degrees of sensitivity. Plasticity of behavior has been neatly categorized by psychologists and includes such phenomena as habituation and various forms of conditioning. The neural bases of such plastic behavior have received extensive study by many workers. In preparations of Aplysia, changes in the size of post-synaptic potentials have been observed during habituation (Kandel and Tauc, 1965). In other preparations, changes in single neuron firing patterns have been observed during various sorts of behavioral conditioning (see reviews by Kandel and Spencer (1968) and Rabinovich (1971). Such experiments have included both intact and *in vitro* preparations, and they have involved the use of appropriate sensory stimuli or, in some cases, electrical stimulation of axon bundles or individual neurons.

A serious difficulty has occurred in such investigations. At one extreme, the animal is chosen to be extremely simple, so that a thorough neurophysiological study is possible, but only limited behavioral plasticity can be attained, as in Aplysia. At the other extreme, rats, cats, rabbits, and monkeys are chosen as experimental subjects. Here there is no problem in obtaining a rich repertoire of behavioral plasticity, but detailed neurophysiological study is impossible. The changes that are observed at the neuronal level are the results of complicated interactions between huge populations of neurons. It is therefore difficult to know whether observed neurons are themselves sources of plastic change or are simply responding as followers to changes in the incident excitation.

Thus a reasonable strategy for the study of plasticity is to find a preparation with a simple nervous system that is yet capable of clearly plastic behavior. In our search for such a preparation, we settled on the crayfish. An extensive literature exists for the anatomy, physiology, and behavior of this animal.

The particular behavior we shall discuss here involves stimulation and movements of the great claws, and is in some respects reminiscent of the cockroach leg experiments of Horridge (1962). The claws are sensitive instruments for the crayfish, and they are particularly visible in Figure 1. They serve as weapons, jaws, and grapples; they are used in exploration and in friendly and unfriendly encounters. Much is known about the claw innervation (van Harreveld and Wiersma, 1937), reflexes (Bush, 1963), and neural discharge patterns (Wilson and Davis, 1965; Smith, 1971).

The large claws are under control of the first thoracic ganglion which contains some 2,000 cells on each side, and is largely unexplored. Although this seems like a large number of neurons, it is likely that only a small number are actually involved with the claw reflexes. The particular movements that we shall discuss involve only two muscles— an opener and a closer. These muscles are controlled by five motor neurons, but it is likely that for the slow movements used here only four of these motor neurons are involved. Thus systems of this sort in the higher crustaceans are probably simple enough for neurophysiological analysis, and they may serve as appropriate tools for the study of plasticity. Other, perhaps more suitable, behavioral paradigms probably can be developed, particularly in conjunction with the well-studied abdominal ganglia (see the review of Kennedy, et al., 1969). The following account is therefore meant only as an introduction of the crayfish into the study of neural plasticity.

Behavioral Methods and Results

Using rubber bands, a healthy, reactive crayfish is strapped, dorsal surface down, to a bar. A small laboratory clamp holds the propodite of one large claw in such a way as not to interfere with movements of the dactyl. The gills are perfused with aerated tap water. A claw position monitoring device is attached. The front part of the head and the eyes are covered with an opaque structure. Under these conditions, the crayfish has no visual input and is free to move only the dactyl of one leg. After a period of adjustment, normal opening and closing reflexes can be elicited; rubbing sensory hairs on the outside of the propodite causes opening, while rubbing sensory hairs on the inside of the claw aperture causes closing. These hair beds are seen in Figure 1; each papilla or pit contains several sensory hairs. Sensory neurons near these structures send their axons into the central ganglia.

We have used a behavioral paradigm with the following steps: 1) Reflex opening is obtained by rubbing the outside of the propodite. 2) A solid barrier is inserted between the dactyl and propodite. This does not touch the propodite, but is in the path of the dactyl if it closes.

FIGURE 1. General view of crayfish claw. Note sensory hair beds inside the claw opening as well as on outer surfaces. Stimulation of the inside hair beds usually results in reflex closing: stimulation of the outside hair beds usually results in reflex opening.

The barrier position may either be inside or outside the normal resting position of the dactyl. 3) Reflex closing is obtained by rubbing the hair bed on the inside surface of the propodite (i.e., within the "jaws"). The dactyl is not touched during this stimulation. 4) When the dactyl hits the barrier during this reflex closing, the dactyl is given a brief opening flick by the same tool that is used to rub the propodite. 5) Stimulation of the inner propodite hair bed is continued for about ten seconds.

The cycle (1) to (5) is repeated every 30 seconds. Ten repetitions constitute a TRAINING sequence. Then: 6) Reflex opening is obtained by rubbing the outside of the propodite. 7) The barrier is removed. 8) Reflex closing is obtained by rubbing the inside hair bed of the propodite. Stimulation continues for about ten seconds. Position of the dactyl is recorded throughout.

The cycle (6) to (8) is repeated every 30 seconds. Two to four repetitions constitute a TEST sequence. Training and test sequences may be alternated and occasional rest periods may be interspaced until either the crayfish or the experimenter gives up. In about 30 percent of the animals so treated, there is some form of "position learning" so that the dactyl eventually closes only to the position where it had hit the barrier during the training sequence. This may be retained for some five or more minutes; retention is usually terminated by a fit of hyperactivity during which the crayfish attempts to move every limb.

Some of the variations of this claw-closing behavior, both during training and test sequences, are illustrated in Figure 2. Stimulation is indicated below each dactyl position record, and the presence and position of the barrier is indicated by the heavy horizontal bar. The top record shows responses of a naive crayfish during his first exposure to the paradigm. The barrier was arranged to be inside the normal resting position of the dactyl. During the test sequence, the dactyl closes smoothly through the position that had been occupied by the barrier.

The middle record in Figure 2 shows the behavior of the same animal after several more training and test sequences. There is now a change in velocity in the claw closure during test as the dactyl passes through the point where the barrier had been. These velocity changes are indicated by arrows.

The bottom record in Figure 2 shows the behavior of another animal which attained very accurate "position learning." The barrier in this case was outside the normal resting position of the dactyl; this accounts for the somewhat different appearance of the movement record. Throughout the whole test sequence shown, the dactyl closed

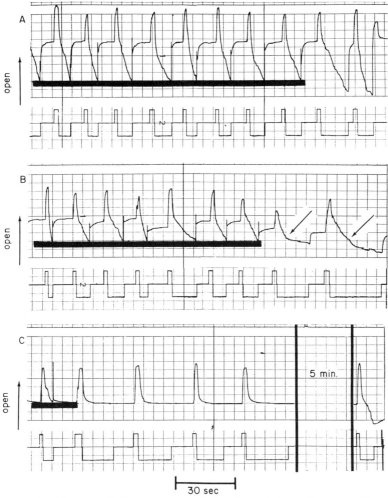

FIGURE 2. Varieties of claw movement during the training and test paradigm described in the text. Upper trace for each record represents claw movement; lower trace represents stimulation of outer (up) and inner (down) hair beds. Presence and position of the barrier during training is indicated by the heavy black bar in the movement record. (A) Naive animal in first training and test session. Barrier is inside the normal resting position of the claw. Upon test, the claw moves through the former barrier position. (B) Same animal during a subsequent training and test session. Note changes of movement at former barrier position as indicated by the two arrows. (C) Another animal. Barrier was outside the normal claw resting position. The last training sequence and four tests are shown. Claw closes only to former barrier position. Five minutes later the claw closes through this position.

only to the position where the barrier had been. As indicated at the extreme right of the record, five minutes later the dactyl closed much further in the same circumstances.

In this preparation, the ordinary claw reflexes of the crayfish are seen to be capable of a certain amount of plastic change. Such changes are somewhat more complex than ordinary habituation, since reflex opening and closing continue to be easily elicited even when the final dactyl position has been "learned." Unfortunately, this is an excessively complicated behavioral paradigm and not a very stable preparation. Nevertheless, it involves a sensory-motor reflex that is limited to a single ganglion and seems worth detailed neurophysiological study.

Electrophysiological Methods

Since the first thoracic ganglion remains largely unmapped (Smith, 1971), we chose to record both from motor and sensory axons near the periphery. This was accomplished through relatively small holes in the base of the propodite which allowed visualization of the motor axons near their entry into the opener and closer muscles. Very flexible suction electrodes were made by drawing Tygon tubing in a small flame. A thin silver wire (10 μ) passed down the suction tubing and protruded very slightly from its tip. Gentle aspiration attached such an electrode to the side of a small nerve bundle and allowed stable recordings without block of the axon even during fairly violent movements of the preparation. After appropriate amplification these spike trains, together with a voltage that represented claw position, were recorded on magnetic tape. These data were subsequently photographed and, where appropriate, analyzed with the help of a computer.

Methods of Analysis

Our general objective is to relate the temporal characteristics of the recorded spike trains to particular stimulus events or to particular movement events. If we are dealing with data from only a single neuron, the appropriate computation is the peri-stimulus time (PST) histogram. This is simply the averaged probability of firing for the neuron as a function of time after some event (the stimulus, for example) (Gerstein and Kiang, 1960; Perkel, Gerstein, and Moore, 1967). Many repetitions of the stimulus presentation are generally necessary in order to obtain a reasonably smooth histogram that closely estimates the underlying firing probability.

If, however, data from several simultaneously recorded neurons are available, we may in addition examine timing relations between the

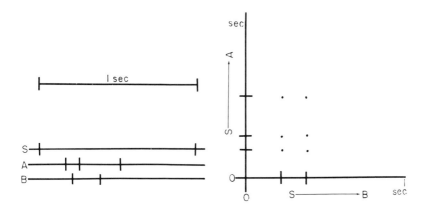

FIGURE 3. Method used to construct joint peri-stimulus-time scatter diagrams. See text (from Gerstein and Perkel, 1971).

observed spike trains and at least partially determine the connectivity or "wiring diagram." The appropriate computation for this purpose is the joint peri-stimulus-time scatter diagram (Gerstein and Perkel, 1969; Gerstein and Perkel, 1971). This diagram is constructed in the following manner, as shown in Figure 3. Let S represent the series of stimulus events, A the series of spikes from one neuron, and B the series of spikes from another neuron. Such data are indicated schematically for a single stimulus interval at the left of the figure. Using the stimulus event as time origin, we mark the several S→A times along the ordinate, and the several S→B times along the abscissa. Points are then plotted on the scatter diagram at all intersections of lines parallel to the axes and through the S→A and S→B times that have been laid off on the axes. All of these steps are indicated in Figure 3. It should be noted that a given A spike gives rise to as many points as there are B spikes in that particular stimulus interval. The same procedure is carried out from each repetition of the stimulus, so that gradually the entire plane fills up with points.

The scatter diagram has a characteristic appearance for each of the possible effective connections to and between the observed neurons. For example, suppose that neither observed neuron is affected by the stimulus, that there is no connection between the two neurons, and finally that they have no shared input. In this case the scatter diagram will show a uniform distribution of points with only statistical fluctuations of point density. Alternatively, suppose that the stimulus causes excitation of the A neuron with some definite latency. Then the scatter diagram will show a band of increased point density parallel to the S→B axis (abscissa) and at a distance from it that corresponds to the latency. Inhibitory influences are made evident as bands of decreased point density.

If the two observed neurons receive shared excitation, or if they have direct synaptic connection, they will tend to fire at fixed time relative to each other. In this situation diagonal bands of altered point density will appear in the scatter diagram at some distance from the principal diagonal; the distance corresponds to the latency of the connections. Variations of point density *along* such a diagonal band mean that the stimulus is able to modify the interaction or shared input. An example, drawn from computer simulation studies, of all these effects is shown in Figure 4. The several marginal densities associated with the joint peri-stimulus-time scatter diagram are shown also as the two PST histograms and the cross correlation histogram.

Electrophysiological Results

We have recorded from various sensory hair axons and from three of the four relevant slow motor axons during both spontaneous movements and during the behavioral paradigm that is described above.

Even with minimal mechanical stimulation of a single sensory hair, it was often possible to record a burst of activity in the appropriate sensory axon followed some 20 to 30 msec later by a burst of firing in the motor axon. Much of this delay can be accounted for by the known slow velocity of conduction. Nevertheless, no direct connections between these sensory neurons and the motor neurons are detected by the methods for spike train analysis that are described above. It is therefore likely that at least one interneuron intervenes between sensory and motor neurons in the claw reflex system; a large amount of sensory convergence does not seem to be needed for this reflex.

We then proceeded to analyze the motor axon discharges. Motor axon recordings were made from the *closer exciter* (CE), the *opener exciter* (OE), and the *opener inhibitor* (OI). Timing relations between such axon firings, the muscle action potentials, and movement of the claw have been previously described by Wilson and Davis (1965). There is a nonlinear dependence of muscle tension on temporal pattern in the firing of the motor axons, as demonstrated by Wiersma and Adams (1950). In addition Wilson and Larimer (1968) have shown that the crayfish muscle is subject to complex hysteresis effects and that the relation between axon firing frequency and tension depends on the recent history of the muscle. Thus the relations between movements, claw position, claw rigidity, and the underlying neural activity are neither simple nor constant.

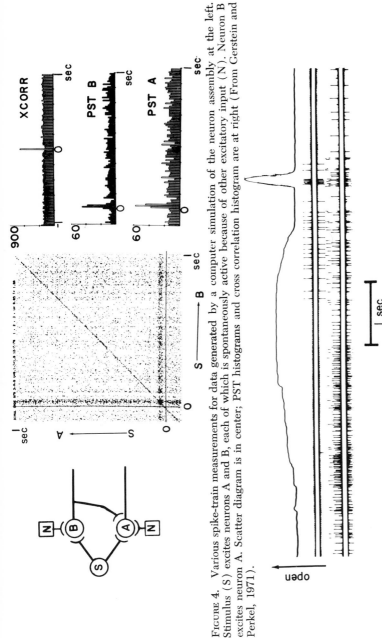

FIGURE 4.   Various spike-train measurements for data generated by a computer simulation of the neuron assembly at the left. Stimulus (S) excites neurons A and B, each of which is spontaneously active because of other excitatory input (N). Neuron B excites neuron A. Scatter diagram is in center; PST histograms and cross correlation histogram are at right (From Gerstein and Perkel, 1971).

FIGURE 5.   Motor axon activities and claw movements during a single iteration of the training paradigm. Upper trace indicates claw position. Middle trace is closer exciter (CE) axon. Bottom trace is opener exciter (OE, small spike) and opener inhibitor (OI, large spike).

A film strip of motor axon firings during a single iteration of our training paradigm is shown in Figure 5. Note that the opener exciter is most active during opening phases of the movement, while the closer exciter and opener inhibitor are most active during closing phases of the movement. Nothing dramatic occurs as the dactyl hits the barrier. All three axons show a burst of activity during the opening flick (step 4 of the behavioral paradigm).

In order to make a more quantitative examination of these firing patterns relative to the movements, we make use of the PST histogram. There is a serious difficulty here, however. The PST histogram calculation assumes that the time structure of each successive repetition of the experiment is identical. In these experiments we begin "opening" stimulation every 30 seconds, but the ensuing movements are not stereotyped either with respect to amplitude or duration as can be seen in Figure 2. Thus the averaging process implicit in the calculation smears out the time structure of the nonidentical successive repetitions. A more correct and accurate way to use the PST histogram and PST scatter diagram in this situation would be to put time markers at various particular phases of each movement, thus creating more appropriate time origins for calculation. It should be emphasized that this was not done in the work reported here. Time origins were taken at the beginning of the "opening" stimulation and a long time base (15 sec) was used.

PST histograms for the activities of three motor axons simultaneously recorded from one animal during both training and test sessions are shown in Figure 6. The vertical scales of the test and training histograms have been normalized to give roughly the same size histograms in spite of a different number of repetitions for the test and training sequences. The statistical variability of activity during the tests appears larger only because there were fewer repetitions. This fiigure shows that for each axon the general shape of the PST histogram, and hence the temporal firing pattern, is approximately the same in both training and test situations.

The PST histograms of Figure 6 showed us the time patterns with which the several axons fired and the average relation of this pattern with the stimuli and claw movements. Now let us seek correlation of firing between two axons. Figure 7 shows PST scatter diagrams for the various possible motor neuron pairs, using the same conventions of time origin at beginning of "opening" stimulation, and the same 15-second time scale. The left half of the figure is from data obtained during the training situation; the right half of the figure is from data obtained during the test situation.

The scatter diagrams calculated from the various axon pairs are different in their general appearance in correspondence to the various different temporal activities of each axon within the open-close movement cycle. However, for each particular pair of axons, the scatter diagrams for the training situation and for the test situation are very similar. Statistical differences again are noticeable because there were fewer repetitions of the test sequence than of the training sequence.

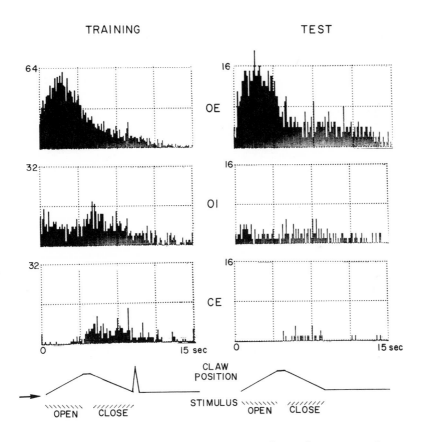

FIGURE 6.   PST histograms of motor axon activities during the training and test sequences. Time origin for each histogram is at beginning of stimulation of outer hair beds. Left column, training ($n = 10$); right column, test ($n = 4$). Axons are labeled in center. Claw position and stimulation are schematically indicated at bottom. The barrier is present only during training and is at the position indicated by the arrow.

Now let us inspect these scatter diagrams for signs of diagonal bands which would, as we have seen, indicate possible shared input to or possible connection between the two neurons under examination. In only one scatter diagram, that for OI, CE (lower left) is there evidence for an excess diagonal concentration of points (the contrast has suffered in the photographic reproduction process; the diagonal band is far more visible in the original or if higher time resolution is used). This diagonal concentration is apparent only in the portion of the scatter diagram that corresponds approximately to the closing movement; it is not seen at earlier or later phases of the movement sequence. Although this calculation of the scatter diagram could be done with more meaningful time origin, as discussed above, these results suggest that the *Closer Exciter* neuron and the *Opener Inhibitor* neuron have a higher than chance probability for nearly coincident firing during the early portions of the closing movement. From the appearance of the scatter diagram, it is likely that this tendency for near coincidence of firing is the result of shared input to the two neurons.

It is interesting to note that the diagonal band is absent in the corresponding test situation (lower right). This might be statistical artifact, but it is worth further investigation. The test situation differs from the training situation only in the removal of the barrier. Thus during a test when the dactyl reaches the former barrier position there is no additional stimulation of the inner hair bed by the barrier itself or by the opening flick by the experimenter.

Thus, from the electrophysiological measurement of spike trains at the periphery, we have inferred the existence of some of the central connections between and to the observed motor neurons. Further details of these connections can be analyzed from Figure 8. Here we have recalculated the scatter diagram for the OI, CE neurons during the training situation. Now, however, we have used a time marker that is completely unrelated to the movement and a short time base of 250 msec. This allows us to examine the diagonal structure with much more resolution although without reference to particular phases of the movement sequence. Essentially this is equivalent to a calculation of the cross correlation histogram of the two spike trains. Clearly there is a band of increased point density that falls on the principal diagonal of the scatter diagram. The general appearance of this diagonal point density continues to support the interpretation that there is shared input to the two motor neurons.

Upon closer examination of Figure 8 (particularly upon sighting along the diagonal), we note that the concentration of points below

TRAINING   TEST

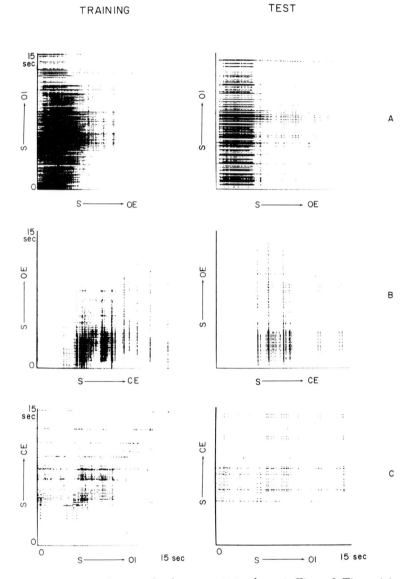

FIGURE 7.   Scatter diagrams for the same activity shown in Figure 6. Time origin
for each scatter diagram is at beginning of stimulation of outer hair beds. Left
column, training; right column, test. Axons are labeled along coordinate axes
of each scatter diagram. See text.

and to the right of the diagonal is considerably lower than to the left and above the diagonal. Point densities far from the diagonal in both directions are equal. The physiological interpretation is not unique, unfortunately, unless additional data are made available. However, a likely candidate "circuit diagram" would suggest that the shared input to the observed neurons is from two separate "follower" interneurons which are themselves driven from a common source. Between the "follower" neurons there may be an inhibitory connection. The common source depends on stimulation of the inner hair bed and on the closing movement. Further details of the "circuit diagram" as well as its changes during plastic behavior remain to be worked out.

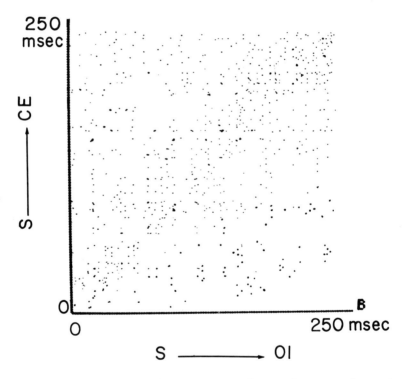

FIGURE 8.   Scatter diagram for the activities of CE and OI axons during the same training sequence as in Figures 6 and 7, but using higher time resolution and a periodic origin time that is unrelated to stimulation or claw movement. This scatter diagram contains only the information of a cross-correlation histogram of the two spike trains.

Conclusion

This paper has presented a new preparation that may be appropriate for the study of the neural changes that underlie plastic behavior. As we have seen, the claw preparation is relatively simple and neurophysiologically accessible, although the behavioral paradigm we have used is complex. A number of additional experimental approaches must be used to extend the work reported here. It is necessary to explore the first thoracic ganglion directly with intracellular electrodes. The first task here will be to locate the cell bodies and neurites of the motor neurons we are interested in. Only the *Opener Exciter* cell body has been found so far (Smith, 1971). Using signal averaging techniques and injection of Procion Yellow marker dye, systematic identification of sensory neurons, interneurons, and motor neurons can proceed.

It is essential to obtain a fairly complete neurophysiological understanding of the claw reflexes before it will become possible to clarify the nature of the changes underlying the observed behavioral plasticity. As we have seen, spike train analysis from identifiable axons near the periphery is just a start on this problem.

In a more general sense, we have tried to indicate the potential utility of the higher crustaceans for studies of plasticity. Interesting and complex behavior can be found in such preparations, and the subsequent neurophysiological analysis is quite feasible.

ACKNOWLEDGMENT: We are grateful to D. Spector and K. Subramanian for computer programs. This work was supported by NIH Grant NS 05606.

Literature Cited

Agranoff, B. This volume, pp. 1-9.

Bush, B. M. H. 1963. A comparative study of certain limb reflexes in Decapod crustaceans. Comp. Biochem. Physiol., *10*:273-290.

Gerstein, G. L., and N. Y. -S. Kiang. 1960. An approach to the quantitative analysis of electrophysiciological data from single neurons. Biophys. J., *1*:15-28.

Gerstein, G. L., and D. H. Perkel. 1969. Simultaneously recorded trains of action potentials: analysis and functional interpretation. Science, *164*:828-830.

Gerstein, G. L., and D. H. Perkel. 1972. Mutual temporal relationships among neuronal spike trains. Biophys. J., *12*:453-473.

Horridge, G. A. 1962. Learning of leg position by the ventral nerve cord in headless insects. Proc. Roy. Soc. (London) Ser. B., *157*:33-52.

Kandel, E. R., and W. A. Spencer. 1968. Cellular neurophysiological approaches in the study of learning. Physiological Reviews, *48*:65-134.

Kandel, E. R., and L. Tauc. 1965. Mechanisms of heterosynaptic facilitation in the giant cell of the abdominal ganglion of *Aplysia depilans*. J. Physiol. (London), *181*:28-47.

Kennedy, D., A. I. Selverston, and M. P. Remler. 1969. Analysis of restricted neural networks. Science, *164*:1488-1496.

Perkel, D. H., G. L. Gerstein, and G. P. Moore. 1967. Neuronal spike trains and stochastic point processes. 1. The single spike train. Biophys. J., *7*:391-418.

Rabinovich, M. Y. 1971. Neural mechanisms of the conditioned reflex. In *Physiology of Higher Nervous Activity*, Part 2, pp. 3-33. Moscow: Nauka.

Smith, D. O. 1971. *Motor Control of the Crayfish Claw*. Thesis, Stanford University, Department of Biological Sciences.

van Harreveld, A., and C. A. G. Wiersma. 1937. The triple innervation of crayfish muscle and its function in contraction and inhibition. J. Exp. Biol., *14*:448-461.

Wiersma, C. A. G., and R. T. Adams. 1950. The influence of nerve impulse sequence on the contractions of different crustacean muscles. Physiol. Comp., *2*:20-33.

Wilson, D. M., and W. J. Davis. 1965. Nerve impulse patterns and reflex control in the motor system of the crayfish claw. J. Exp. Biol., *43*:193-210.

Wilson, D. M., and J. L. Larimer. 1968. The catch property of ordinary muscle. Proc. Nat. Acad. Sci., *61*:909-916.

Development and Patterning of Movement Sequences in Inbred Mice

JOHN C. FENTRESS
Departments of Biology and Psychology
University of Oregon
Eugene, Oregon

THE TEMPORAL DISTRIBUTION of movement patterns represents a fundamental and enduring problem in behavioral biology. Animal species interact with their environment largely through the articulation of species-characteristic behavior sequences, a fact recognized a hundred years ago by Charles Darwin and pursued in earnest at the present time by ethologists and related workers (for example, Darwin, 1872; Hinde and Stevenson, 1969). It is apparent that the central nervous system itself operates in important respects as a temporal machine devoted to the adaptive patterning of movement sequences (Lashley, 1951; Sperry, 1952; Evarts, 1971), but our understanding of both behavioral rules and temporal patterns of neural activity is at present at an elementary stage.

In a recent series of articles Hinde has stressed some of the major issues in the control of animal movement patterns which are relevant to the present study (Hinde, 1969; Hinde and Stevenson, 1969; Hinde and Stevenson, 1970). The first problem concerns precise description of the stream of events that we call behavior. One implication of emphasis upon behavior as a temporal stream is that it is often valuable to examine a given behavior within the context of preceding, subsequent, and alternative activities. How, for example, are

83

different component movements linked together over time in an unrestrained organism? Interactions between behaviors can provide important clues about the controls of a given behavior since regularities in temporal patterning can suggest causal relations between these activities. How separate are the control processes that underlie different behaviors; what are the rules which string these behaviors together over time? Secondly, behavior represents a subtle interplay between events originating within the organism (endogenous) and those originating in the outside world (exogenous), and the types of interaction between these endogenous and exogenous influences are of obvious interest. At a slightly more analytical level, the question may be phrased in terms of interactions between central and peripheral mechanisms. Thirdly, analysis of these integrative functions that contribute to organized patterns of behavior can be usefully supplemented by some consideration of stored information. This in turn is a question of behavioral development (ontogeny) in which *both* genetic and experimental sources play a fundamental role.

The present report concerns aspects of the temporal distribution, control, and development of movement patterns in mice. Attention will be concentrated upon the organization of grooming sequences, but these behaviors are first briefly examined within the context of other ongoing activities. Grooming is of particular interest for several reasons: a) It is an often repeated behavior that occurs under predictable circumstances and in rather close association with certain transitions in other activities; b) it is a relatively stereotyped yet complex collection of movement patterns, making it possible to subdivide into components and to seek rules of temporal connection between these components; c) it can on occasion be enhanced by stimuli that do not appear immediately relevant, and as such may provide an assay of problems of specificity of integrative control in patterned behavior; d) questions of peripheral versus central control are obvious and can be pursued with some degree of precision; and e) components of grooming develop early in mice and show certain strain characteristic patterns under standardized environmental conditions. The data not only relate to previous studies of grooming in other rodent species (Fentress, 1968a, b), but they may provide clues with respect to thinking about other temporal patterns of behavior.

The fundamental nature of problems of serial organization of behavior can be seen, for example, in human language, and it has recently been suggested by several workers (e.g., Lenneberg, 1967; Vowles, 1970) that there may be some common principles of organization among human language and the movement patterns of animals. This is at present, of course, a point of view based upon often rudimentary analogies, but it does emphasize the potential excitement that can come from detailed descriptions, classifications, and analyses of movement patterns in animals. At present even our descriptive data on mammalian behavior patterns are inadequate to be any more than suggestive (cf. Fentress, 1967).

Ongoing Behavior Patterns in Mice

Most of our initial observations on mice are made by placing an individual animal in a relatively large enclosure (e.g., 2 feet by 4 feet) and recording *continual* sequences of the animal's activity patterns. In recent months we have developed a manually operated recording tone box which inaudibly records behaviors onto tape for subsequent computer analyses (Linc 8). One such record is shown in Figure 1. Ongoing behavior was subdivided into ten categories which may be reliably distinguished from one another. The four lines at the top of the figure indicate sequences of behavior as they occurred during the first ten minutes after the animal was placed in the observation box. Major divisions of grooming (face, belly, back, tail) are in bold type. Their distribution is characteristic and indicates some interesting features of sequential organization. Mice typically spend the first few moments in the observation box moving about and pausing. Brief episodes of face grooming are interjected, in turn followed by pauses and subsequent locomotion.

Later in the trial, grooming sequences become more prolonged. They again start with the face, but then they progress without interruption to the belly and then the back. Still later the animals show extended periods of vigorous grooming in which the movements may go directly from the face to the back, often followed by grooming of the tail. This means that the organization of movement sequences that include grooming components is not constant, but more interestingly shifts in a systematic manner during the first few moments of the trial. There appear to be different levels of control, some concerned with the patterning of grooming components themselves and others concerned with the temporal linkage of these components with each other and surrounding behaviors. The sequential linkage of individual components of behavior can be modulated by control processes which shift in a systematic manner through the early period of the trial. We shall return to this point again.

To plot individual behaviors with respect to immediately preceding activities is of interest. This provides an overall summary of the sequential linkage between different activity patterns. This is done in the plot on the left of Figure 1 under the title SORTED ARRAY H2. Preceding behaviors are listed along the left and subsequent behaviors across the top. Thus if we look at B (locomotion) as a preceding activity (using the column at the left), we find that it directly precedes A (inactivity) 15 times during the 10-minute period. The distribution of numbers in the cells thus provides a descriptive summary of the degree of sequential linkage between any two behaviors. Information measures may be applied to such a table to estimate the predictability of a given behavior as a function of its overall probability and direct sequential connection with other activities (e.g., Attneave, 1959; Garner, 1962).

Notice that a behavior never precedes itself directly in the plot. This is because for the present purposes bouts are treated as unitary events, and the symbol is only changed when the next behavior occurs. Thus an uninterrupted period of walking is entered once. The alternative method is to record one behavior at set time intervals (e.g., once per second), but this results either in many consecutive repeats of the same symbol when short intervals are used or misplaced sequences of momentary behaviors when long intervals are used. For example, if walking occurs for ten seconds and we record once per second we obtain, in one walking sequence, nine entries of walking preceded by walking (the first occurrence of walking is by definition preceded by something other than walking). If ten-second recording intervals were used, brief occurrences of a given behavior might be missed entirely. Such data would be less interesting for a mathematical summary of sequential coupling *between different* activities (cf. Chatfield and Lemon, 1970).

The information measures used are based upon the transformation of numbers to a logarithmic scale with the base 2 for convenience. Thus the number 3 in this nomenclature represents eight (equally probable) events $(2^3 = 8)$, and 3.16993 represents the nine behaviors this animal actually showed during this part of the trial. This is called HO. If the animal showed each behavior equally often, and these behaviors were sequentially independent, our chances of guessing its behavior at any particular instant is represented by HO and equals one out of nine.

It is apparent that the animal shows some behaviors much more often than others. This makes the situation rather like betting on a weighted coin, and it improves our ability to guess what the animal will be doing at any instant. This weighted probability of behaviors is represented by H1. For this mouse, H1 is approximately 2.7. This means that if we were informed of the different probabilities of the different activities, we would guess the mouse's behavior on the average

Temporal Patterns of Behavior

Successive Behaviors observed (see key) with grooming behaviors in bold type:

```
A B A F A C A B A B A B A B A E A F A B A B A B A F A B A C E A E A F
A E A B A F A B A E E B E B E A B B E A B B E A F G H A E F H I
F H I H F I F A F A E E A B E J A B A B E B E B A B C B A
B A E E B B E A F E A F H G F G F E B B E A
```

Behavior displayed during first ten minutes

	Sorted Array H2												Sorted Array H3°									
Behavior preceded by	A	B	C	D	E	F	G	H	I	J		A	B	C	D	E	F	G	H	I	J
Key																					
A—Inactive	17	3		8	10							21	1			7	1	2	1		1
B—Locomotion	15	1		11	2							4			4	4					
C—Chew	1	1										3			1						
D—Dig																					
E—Stand	11	9		1			1	3	4			5	3			7	4	1	1		1
F—Groom face	8				1		4	1				1	4			3	4		2		
G—Groom belly						1	2			2		1	1			1	1			1	
H—Groom back	1					1	2									1	3			2	3
I—Groom tail						2				2		1				1	1		1		
J—Jump	1						1									1					1

Information calculations

HO=3.16993 [1/9]
H1=2.66507 [1/6-7]
H2=1.60480 [1/3-4]
H3°=2.08559 [1/4-5]

HO—H1=0.50486
H1—H2=1.06027
H1—H3°=0.57948

HO—H2=1.56513
HO—H3°=1.08434

FIGURE 1. Temporal patterns of behavior in DBA-wild hybrid *Mus musculus* displayed during first 10 minutes after placement in observation cage. Top lines represent successive behaviors observed (see key). Grooming behaviors are in bold type. H2 sorted array displays number of times each current behavior (columns) was preceded directly by each other behavior (rows). H3° sorted array displays linkage between each behavior and the behaviors two links earlier in the sequence. Information calculations are given in the lower right-hand section of the figure. Bracketed numbers indicate average probability of correct guess of behavior displayed at any instant of time. (See the text for further details.)

of one out of six to one out of seven times. The interesting question is, "what if we also know the animal's preceding behavior?" This is represented by H2 which in this case equals 1.6. Knowledge of the preceding behavior permits us to guess correctly between one out of three to one out of four times. This in turn gives us a feel for the tightness of sequential coupling among the recorded components of behavior.

Obviously, mammalian behaviors are neither distributed randomly with respect to one another nor do they occur in invariable sequences. Probability rules of this type are important in our thoughts about behavioral (and neural) organization, and go under the label *stochastic*. (Previous analyses of these functions in ongoing sequences of animal behavior have been reported by Nelson, 1964, Fentress, 1965, and Altman, 1965.) If we look closely at the H2 plot in Figure 1 we notice another interesting feature of organization which is reflected in many types of behavior, its symmetry. That is, A precedes B often and B also often precedes A; G commonly precedes F and F commonly precedes G. That means that, in addition to going through sequences of behavior, animals tend to oscillate between two behaviors which is in a sense the antithesis of sequences involving more than two activities. This balance between progression through a series of behaviors and oscillation between two behaviors is interesting to contemplate because it occurs commonly and has not been adequately studied.

One suggestion is that the tendency to perform a behavior may persist for some period of time after the behavior is interrupted and replaced by another activity. Independent support for this suggestion was obtained by the author in a previous study (Fentress, 1968a, b). Voles that were either locomoting at the instant that an overhead moving object was presented, or which *had been* locomoting within the previous 10 seconds, were much more likely to flee than animals that were doing anything else, and did not differ significantly from each other. This suggests that some underlying processes active during locomotion persisted for a period of time after locomotion itself had ceased, and that it was at least partially the action of these processes that contributed to fleeing behavior. Comparable evidence has been obtained through direct electrical stimulation of neural loci (e.g., von Holst and von St. Paul, 1963; Fentress, unpublished observations). Related issues are discussed by Hinde (1970).

The plot under the label H3* helps demonstrate the tendency of an animal to return to a preceding behavior after it is interrupted by another activity. In this plot a behavior is related not to the immediately preceding activity, but rather to the one activity immediately prior to the preceding behavior. Notice that the diagonal in the plot generally

Temporal Patterns of Behavior

Successive Behaviors observed (see key) with grooming behaviors in bold type:

B A E A E A E A F A E A E A B A B A B A B A B A B A E A E A E A B E B E
B E B E A F B A B E B A B A B E B A B A B A B E B A B A B A B E A A E B
A E B A B A B A E A C B C E A E A

Behavior displayed for the ten minutes subsequent to disturbance

Sorted Array H2

Behavior preceded by — Key	A	B	C	D	E	F	G	H	I	J
A—Inactive	25	3		13		6				
B—Locomotion	26	1		9		3				
C—Chew		1		3						
D—Dig										
E—Stand	18	8				1				
F—Groom face	4	1				1				
G—Groom belly										
H—Groom back										
I—Groom tail										
J—Jump										

Sorted Array H3*

Behavior preceded by — Key	A	B	C	D	E	F	G	H	I	J
A—Inactive	35	4					9			
B—Locomotion	3	24		1			8			
C—Chew	3			1						
D—Dig										
E—Stand	5	7		2			7	4		
F—Groom face	2	2					2	2		
G—Groom belly										
H—Groom back										
I—Groom tail										
J—Jump										

Information calculations

$HO = 2.58496$ [1/6]
$H1 = 1.95465$ [1/3-4]
$H2 = 1.26982$ [1/2-3]
$H3^* = 1.39755$ [1/2-3]

$HO—H1 = 0.63031$
$H1—H2 = 0.68483$
$H1—H3^* = 0.55710$

$HO—H2 = 1.31514$
$HO—H3^* = 1.18741$

FIGURE 2. The same mouse as in Figure 1 after its ongoing behavior was interrupted by overhead moving object 10 minutes into the trial. Behaviors are displayed for the 10 minutes subsequent to this disturbance. Other details are as in Figure 1.

contains the highest numbers; this reflects the tendency of a behavior to reoccur immediately after it is interrupted. H3* in this case gives a predictability of one out of four to five, which is not quite as high as H2 (but recall in H2 a behavior cannot immediately precede itself, which eliminates all entries along the diagonal in the plot).

Two additional points should be briefly mentioned at this stage. The first is that the mathematics employed above to summarize sequential linkage assume that the probability rules which underlie this temporal coupling are stationary or *ergodic;* that is, they do not change. We have already seen evidence which suggests that this is probably not strictly true, for the probability of face grooming followed by body grooming shifts during the trial. This reduces the predictability of overall sequential relationships as assayed by the present method. Secondly, it is possible that the duration of a given act may be used as a predictor of the subsequent behavior. This information also is not contained in the above evaluation. We shall look at each of these points shortly.

It should not be assumed that linkages between successive pairs of behaviors exhaust all possibilities for rules of temporal ordering, although they suffice for our present purposes (cf. Attneave, 1959; Vowles, 1970). Fentress (1965), for example, found that consideration of triplets of vole behavior in all possible combinations as opposed to just pairs produced a marked reduction of uncertainty in prediction. Supplementary methods for examining longer sequences of mouse behavior have recently been reported by Guttman and others (1969). It must also be explicitly recognized that observations of free-running behavior do not in themselves indicate lack of modulation by environmental input (e.g., Hinde and Stevenson, 1970; Vowles, 1970), and it is often valuable to interrupt ongoing sequences of behavior with an abrupt external stimulus (Fentress, 1968a, b).

Figure 2 shows the behavior of the same animal during the next 10 minutes of the trial, after it had been frightened by a sharp and brief movement of an overhead object (labeled a "bird" by Mr. McDonald, the graduate student who designed this particular test). Overall sequence analysis demonstrates that the animal is now more predictable, primarily because of a reduction in the number of different behaviors shown. Note the several occurrences of face grooming, again in bold type in the figure. Figure 3 plots the occurrence of face grooming, as well as locomotion and inactivity, over time. The animal first showed a long period of inactivity ("freezing"), then groomed, then became active, then groomed, and then settled into a period of less activity. Data such as these suggest the possibility that grooming

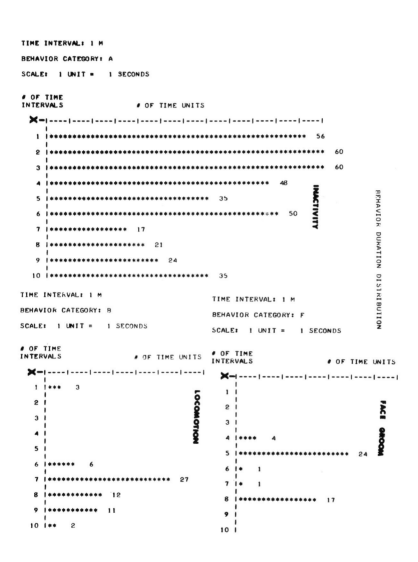

FIGURE 3. Inactivity, face-grooming, and locomotor data from Figure 2 plotted in terms of number of seconds per successive one-minute intervals following the overhead disturbance. Note temporal interrelations between behaviors.

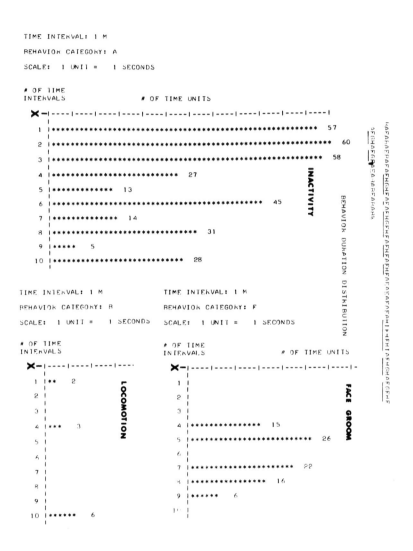

FIGURE 4.   Data obtained from second DBA-wild hybrid mouse after presentation of overhead disturbance 10 minutes into trial. Top lines represent sequence
of behaviors as in Figures 1 and 2. Graphs for inactivity, face grooming, and
locomotion as in Figure 3. Note increase then decrease in grooming after inactivity as in Figure 3, but also lack of subsequent locomotion (cf. Fig. 3).

may occur at a predictable time with respect to an external disturbance, and also with respect to other behaviors.

The results plotted above are reliable and support previous data obtained with voles (Figure 6). Some variations in the overall pattern can often be observed, however. Figure 4 shows data obtained from a second mouse tested as above. Again the animal initially freezes when the stimulus is presented (though not quite as severely as the former animal) and shows a subsequent peak and then decline in face grooming. In this case, however, grooming is not followed by locomotion. (It was followed by considerable standing, not shown.) The point to be made here is that grooming probability appears to be more precisely connected with the stimulus presentation (and intervening freezing behavior) than with any one subsequent activity (cf. Fentress, 1968a, b). Figure 5 shows an animal that gave only a weak response to the experimental stimulus. The interruption of locomotion was momentary, and the grooming that followed was of short duration. We have also observed animals that freeze for the remainder of the trial after the stimulus presentation. Obviously under these conditions all grooming, as well as other activity, is eliminated.

Model of Grooming Integration

Our present data support previous work (Fentress, 1968a, b) that indicated that the amount of grooming can actually be increased over control periods by the presentation of a moderate disturbance. Not only can the duration of grooming be increased, but it may occur in a particularly vigorous fashion after the animal is disturbed. For example, notice the immediate presence of face-to-back sequences in Figure 4. At other times grooming is explosive in its expression but quickly interrupted. This is more the case in Figure 3, and it usually happens when the animal appears to have been disturbed over some optimal level. One might argue that grooming is actively interfered with by other elicited behaviors in these latter circumstances. (A severe disturbance may obviously produce only freezing for the time period recorded, and grooming thus goes down.) Animals that have weak reactions to the test stimulus, as measured by, for example, freezing duration, often show short latency grooming and less total grooming (Fig. 5). Locomotion may or may not follow grooming (compare Figs. 3 and 4).

Grooming occurs under predictable conditions, often those which might be labeled as moderately "stressful," "arousing," and so forth. This predictability assists the investigator who wishes to study in more detail the components of the behavior, and it is also of some interest in its own right. The fact that grooming can be increased, both in apparent vigor of movement and duration, by moderately intense stimuli with which it is not obviously associated is also pertinent. Two major hypotheses that might account for the phenomenon are a) that grooming can be excited (in the behavioral sense) by apparently irrelevant stimuli, at least at certain intensities; and b) that the stimulus suppresses a behavior which normally suppresses grooming and thus grooming is "disinhibited." The former hypothesis suggests that excitatory processes underlying different behaviors may not be as separate as they first appear, but rather that they overlap to varying degrees (analogous to subjective experience of merging bands of color in a rainbow). The second model stresses excitatory specificity, with interactions between behaviors of an inhibitory nature only. This is not the place to go into a detailed evaluation of these alternatives, but is does appear that it may be useful to supplement the latter ("disinhibition hypotheses," e.g., Sevenster, 1961; Rowell, 1961) with the former (excitatory overlap) model (Fentress, 1968a, b). For example, while grooming often occurs in the transition between freezing and subsequent locomotion (each of which might be said to inhibit grooming which is then excited by its own private controls), the transition is variable: total grooming is often greater than would be expected by the disinhibition model without additional assumptions, and the movements themselves often appear unusually vigorous.

For the purpose of discussion (or debate!), Figure 7 shows a sketch of a brief scheme that may have some applicability to the temporal integration of grooming and perhaps other behaviors as well. It concerns the effects of a secondary input upon a predominant and relatively stereotyped behavior such as grooming. A secondary input is defined as one that appears to be primarily associated with a behavior other than the one in question. For example, the "bird" is primary with respect to freezing and secondary with respect to grooming. A predominant behavior refers to one that is well established in the organism's repertoire and has a high probability of occurrence, such as grooming in mice. On this model the secondary input (one that at a higher intensity is primarily associated with a behavior other than the one studied) can facilitate the predominant behavior when the input is of low or moderate intensity. (More critically, physical input intensity should be supplemented by evaluation of the effectiveness of a given stimulus in eliciting primary behavior, which in turn is partially a function of the animal's internal state; however, this approximation will serve our present purpose). At higher input intensity the predominant behavior (e.g., grooming), to which the input (e.g., "bird") is defined as secondary, will be suppressed and probably replaced by the behavior primarily associated with the input (e.g., fleeing/freezing).

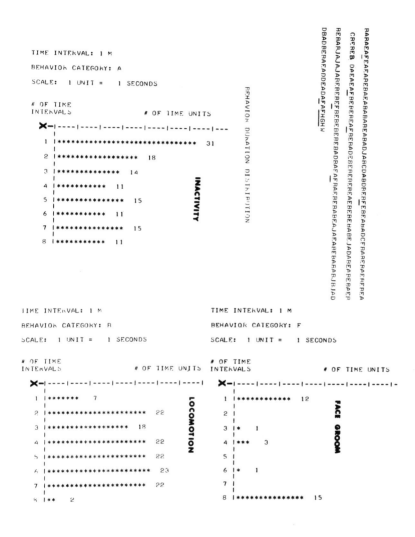

FIGURE 5.   Data from third DBA-wild hybrid mouse that showed only a mild response to overhead disturbance. Procedure and data presentation as in Figures 4 and 5. Note early occurrence but low total grooming and early occurrence of locomotor activity in comparison to Figures 3 and 4.

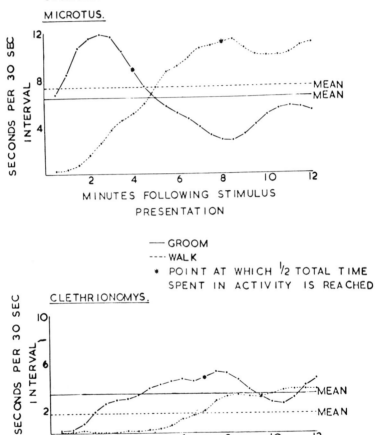

FIGURE 6. Smoothed mean grooming and locomotor behavior in groups of *Microtus agrestis* and *Clethrionomys britannicus* (voles) after presentation of overhead disturbance. Compare to Figures 3-5 and 7. (From Fentress, 1968a.)

MODEL FOR SECONDARY INPUT EFFECTS

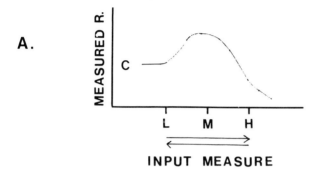

A.

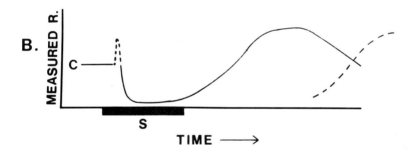

B.

FIGURE 7. General model for effects of secondary input upon predominant (and relatively stereotyped) behavior patterns. A. Overall effects of different input intensities (L = low, M = moderate, H = high) upon response (R) in comparison to control level (C). Model suggests that moderate intensity input can facilitate an apparently irrelevant predominant-stereotyped behavior (e.g., grooming) but that the same input at high intensity may block the same behavior. B. Effect of high intensity input over time during and after a high intensity stimulus (S). Picture suggests that a momentary facilitation may occur upon initial presentation of stimulus (dotted line at the left), but this is soon replaced by suppression of the behavior. "Rebound" may occur after termination of stimulus. Dashed lines at right of Figure represent behavior that subsequently replaces predominant-stereotyped activity (e.g. locomotion subsequent to grooming, as in previous figures). (See the text for details.)

One implication of this model is that control systems underlying different behaviors are neither entirely separate ("specific") nor undifferentiated ("nonspecific"), but rather may under appropriate conditions (e.g., appropriate input intensity) show partial overlap in terms of shared excitatory pathways (defined here by functional rather than anatomical criteria). Such a principle of organization is somewhat more difficult to conceptualize than absolute models of specificity *versus* nonspecificity, but it appears to fit a variety of behavioral observations (*see* Fentress, 1965; Hinde, 1970; and Valenstein, 1969 for general reviews). Such a picture is not only compatible with behavioral data but also with data at the neurological level (Luria, 1966) as well as neurophysiological levels of analysis. With respect to the latter, for example, Evarts and Thach (1969) write, "it would seem likely that gradients rather than sharp boundaries are a general feature of patterns of central organization in the motor system as in other systems: localization is more-or-less rather than all-or-none" (p. 471).

The value of the model for displacement activities (excitation of apparently irrelevant behaviors) was proposed by Fentress (1965, 1968a, b) and recently confirmed by Wilz (1970). The scheme proposed here goes somewhat further and suggests that the degree of positive overlap among behavioral control systems can shift not only as a function of the internal dynamics of the organism, but most explicitly, as a function of stimulus intensity. As the intensity of input is increased, the range of outputs that can be facilitated often appears to decrease; that is, specificity of control pathways as measured by secondarily related outputs increases (grooming is suppressed rather than facilitated). Such a dynamic view is also compatible with recent neurophysiological literature, although it must be emphasized that the levels of analysis are very different. Eccles and others (1967), for example, report that the spread of Purkinje cell activation in the cerebellum by mossy fiber input becomes progressively restricted with increased intensities of the input due to "focusing action" mediated by Golgi cells. Surround inhibition studied extensively in a variety of sensory and motor systems could provide similar increased specification with increased input intensities. It is possible that behavioral control systems interact with one another in an analogous fashion under certain circumstances, although the level of complexity here is obviously much greater.

Part B of Figure 7 summarizes what may happen *during* and *after* a *high intensity* input (right hand part of A). We have found that suppression of ongoing behavior by a stimulus may be momentarily preceded by a brief intensification (e.g., speeding up) of the behavior (the dotted line in the figure), and that *after* the stimulus the suppressed behavior may "rebound" to a level above that of a control period (C). This is, the main excitatory effect of a secondary stimulus at high intensity occurs at the termination of the stimulus, and, within the limits measured to date, the latency of this secondary excitation is a function of the intensity of the primary response to the stimulus (Figs. 3-5). This is the situation most similar to the experiments reported above. The dotted line in Figure 7B represents a subsequent behavior, such as locomotion in the present study. Comparison of the schematic with actual data obtained for grooming in voles in a previous study (Fig. 6) demonstrates quite a close fit. The traditional model for such secondary facilitation ("rebound") is an accumulation of excitation specific to the facilitated behavior during the period that it is blocked. An alternative, or supplement, suggested for consideration here is that the aftereffects of a moderately intense stimulus may result in functionally diffuse excitation that can "feed into" a predominant behavior such as grooming.

Another implication of the scheme presented is that the *very initial* response to a moderately strong stimulus is relatively unspecific, and that this can account for the momentary acceleration of an ongoing behavior represented by the dotted line a the left of Figure 7B.

There is a need for further critical evaluation of the integrative dynamics proposed here. Brief additional examples may clarify the main outlines of the proposed scheme. Fentress (1965) found that persistent locomotor stereotypes, such as seen in many zoo animals, can be facilitated by a particularly wide range of stimuli over a broad spectrum of intensities. The major relations can be most clearly seen as the locomotor stereotypes become increasingly predominant. For example, study of pacing in a Cape hunting dog at the Zoological Society of London revealed that a variety of moderate intensity disturbances, such as approaching food carts or school children, would facilitate this highly predominant behavior, but that at higher intensities and/or closer proximity of the disturbances, the animal's response became more specifically focused with respect to the stimulus. Pacing was suppressed and replaced by specific approach or avoidance behaviors. *After* a high intensity disturbance (e.g., the school children leave), however, pacing was typically facilitated to a marked degree. Sudden disturbances produced momentary accelerations in the pacing behavior, *then* specific oriented reactions. Similar observations have been made with running wheel behavior in our laboratory mice. One additional example from the mice should suffice here.

If one anesthesizes a mouse and watches it during recovery, one typically observes the following sequence of recovery behaviors: a) scratching with one rear leg; b) chewing; and c) face washing. If one gives the mouse a moderate pinch on the tail during the rear leg scratching period, scratching will be facilitated during the pinch itself. This appears to be an irrelevant response since the scratching is oriented toward the ear. A strong pinch, however, suppresses scratching, with occasional momentary facilitation, which then rebounds [after the stimulus ceases (as measured by probability, duration, and/or vigor of the scratching)]. *During* the strong stimulus the animal may make a slight body turn toward the pinched tail (viewed here as a primary response). When the animal is at the chewing stage, a light tail pinch now "feeds into" and facilitates chewing. A strong pinch again suppresses chewing, which may rebound. The same relations hold true for face washing, although here the results are more statistical due to the fact that the animal is nearly awake. In summary, the input (tail pinch) can facilitate a variety of outputs at moderate intensities, but at higher intensities the response is specified to twisting of the body. Momentary facilitation and rebound can be seen with more intense stimuli.

As Hinde has consistently emphasized (e.g., Hinde, 1970), models of behavior such as the above should be taken as approximations to guide future thinking and should not be overly interpreted in terms of their universal or detailed applicability. The above examples were purposely drawn from diverse observations, and the appropriate suggestion is that there are broad similarities, although not necessarily identities, in the functional principles outlined. The scheme outlined was based primarily upon grooming data in freely moving rodents. To what extent it can be extended is of course a matter for further empirical investigation. We already have data that indicate that other factors (e.g., diffuseness versus localizability of the stimulus, number of behavioral alternatives available to the organism) must be considered for a more complete and precise picture. For example, when mice are placed individually in small plastic containers, they give the impression of being quite agitated or "aroused," but have few behavioral alternatives available to them. Grooming and related activities are particularly predominant under these conditions.

Comparison of Grooming in Inbred and Hybrid Strains

The individual records shown in Figures 1-5 were obtained from animals that resulted from a cross between a DBA (inbred) mouse strain and a wild male house mouse. We have found that the grooming in wild house mice is particularly predominant, and thus these animals provided useful data for our preliminary studies. The fact that different inbred strains differ in their tendency to show face grooming under stardardized test conditions is clear from Figure 8. DBA/2J mice show considerable grooming in comparison with C57/6J mice, and F_1 hybrids between these two strains display an intermediate value (as well as greater variability). The three heavy lines in the figure represent the mean values of face-grooming time for the three groups. Individual animals are represented by the thin lines. Each animal received several trials, spaced as indicated on the abcissa.

The data in Figure 8 are of interest not only because they illustrate genetic differences in grooming, but also because they lead to a potential caution in generalizing behavioral data from a single strain. We have found that moderate disturbances can facilitate the occurrence of the relatively predominant grooming in DBA mice and hybrids, but that the lower baseline grooming in C57 mice is less often facilitated by similar disturbances. C57 mice show much digging behavior, and it is this predominant motor activity that is likely to be facilitated by situations that produce increased grooming in DBA, wild, and hybrid mice.

Genetic differences in grooming behavior should not be taken to mean that environmental factors in development are unimportant. For example, Mr. D. McDonald in our laboratory has found that C57 mice reared in large cages and subsequently transferred to small cages become much more reactive to external disturbances than mice reared in small cages. Grooming was also facilitated in the animals reared in large cages. The effects of differential rearing have been found to last over a period of months. Thus, to say that genetic differences produce differences in behavior is quite different from any implication that the behavior is solely under genetic control (cf. Hinde and Tinbergen, 1958; Fentress, 1967, 1968a, b).

Components of Face Grooming

Grooming is not a unitary behavior, but contains a variety of components. We have already seen, for example, that one can subdivide grooming into major segments such as face grooming, belly grooming, back grooming, and tail grooming. Furthermore, we have seen that these divisions of grooming bear a relatively systematic relationship to one another, and that overall grooming occurs under predictable circumstances (i.e., in relationship to environmental disturbances and other behaviors). At this stage it is of interest to describe the major

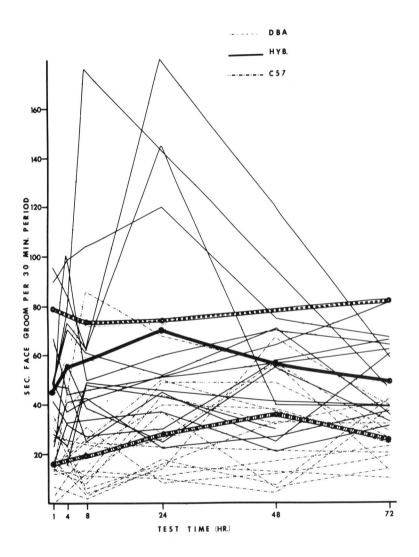

FIGURE 8.   Face grooming in DBA, C57, and DBA/C57 hybrid mice during 30-minute test periods (animals individually placed in plastic cages). Heavy lines represent mean data for each group; thin lines represent data from individual mice. Times of repeated test periods (with reference to first test period) indicated on abcissa.

STROKES

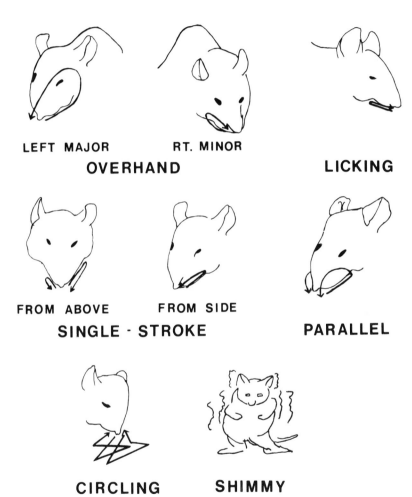

LEFT MAJOR   RT. MINOR

OVERHAND   **LICKING**

FROM ABOVE   FROM SIDE

SINGLE · STROKE   **PARALLEL**

CIRCLING   **SHIMMY**

FIGURE 9.   Major components of face grooming. (See the text for details.)

components of face grooming in greater detail and to seek sequential relationships between these components. The data reported here were obtained from single-frame analysis of 16mm film (64 and 32 f.p.s.) taken of C57 mice. We have related information from other strains, and most overall patterns appear similar, although there are some differences in detail.

We have been able to isolate reliably six active components of face grooming as illustrated in Figure 9. One of these, licking, has been subdivided into long and short bouts to assist sequential analysis. Our analysis also includes pauses since they are momentary interruptions in face-grooming sequences that occur at relatively predictable times during these sequences. A brief description of the major face-grooming components follows.

Overhand strokes consist of a large somewhat circular movement of the major limb from underneath the chin up onto the head and down over the top of the nose. The major limb usually crosses the midline of the head sometime in the trajectory. The opposite limb, or minor limb as we have called it, moves synchronously with the major limb but participates in a much smaller amplitude movement. Overhand movements may either follow one another directly or be separated by other grooming components. Each limb participates in major and minor movements, and during bouts with close successive overhands the limbs alternate with respect to major and minor movements in a relatively strict fashion. The strictness with which a given limb alternates between major and minor movements declines with increasing intervals between successive movements. The duration of overhand movements varies as a function of the distance traveled by the major limb, with a mean duration of approximately one fifth second for C57 mice. During a sequence of overhands successive movements tend to be somewhat longer in distance and duration, although this pattern can vary.

Licking occurs with both hands underneath the snout, usually symmetrically. One licking cycle usually starts at the nostrils or just in front of the nostrils (although very occasionally behind), progresses to the rear edge of the snout, and then forward again. The primary movement is in a horizontal direction. While normally actual licking of the forepaws with the tongue is observed, occasionally position of the animal with respect to the observer prevents direct observation of paw wetting, and to date its occurrence is assumed under these conditions. We have often found it convenient to subdivide licking into bouts of different lengths (e.g., short = less than $\frac{1}{3}$ second, etc.—see below) because licking duration can be a predictor of subsequent behaviors. As far as we can determine the basic licking movement remains constant in bouts of different duration.

Single strokes are basically horizontal movements from the snout as far back as the cheek. As with overhands the two limbs move synchronously but are asymmetrical with respect to distance. Unlike overhands, single strokes do not move forward over the top of the snout. Single strokes may take as long as one third second when they occur separately. More typically, however, a series of single strokes follows another at one-tenth-second intervals. During these repeated strokes the limbs alternate with respect to major and minor amplitude movements.

Parallel strokes are half moon in shape, with both limbs moving simultaneously and symmetrically with respect to the pathway covered. The pathway traveled is slightly more rounded than in single strokes, but not nearly so much as in overhands. Parallel strokes may vary in duration, but they typically occupy slightly more than one tenth of a second.

Circling consists of an irregular and extremely rapid series of forepaw movements in front of, and usually below, the face. The movements occur in both horizontal and vertical vectors, and a given limb may change direction up to 30 times per second. Explosive is perhaps the best description of these brief action sequences. Five or six movements of a given limb may comprise an entire bout.

Shimmy is an explosive shuddering of the entire body. Single frames may be blurred even when taken at 64 f.p.s. Like circling, shimmy does not represent a face-grooming stroke per se, but rather is a characteristic movement pattern that is closely associated with face-grooming strokes.

Pause is self explanatory. The paws usually rest at either the mouth or chest. We have sometimes found it convenient to subdivide face grooming even further, such as *crossing* of the forepaws after an overhand stroke and *descent* of the paws from the snout after a grooming cycle. However, for our present purposes these main categories will suffice.

Sequential Coupling of Face-Grooming Components

Differential probabilities and sequential coupling of face-grooming components are shown in Figure 10. We see that face grooming is far from being a unitary behavior, and that the individual components follow one another in a probabilistic fashion. Notice that in this plot successive overhands are listed. These are large and relatively variable movements that can occur either separately or in strings, and thus it seemed of interest to include this information (an additional reason will become apparent below). Single strokes, by contrast, almost always occur in strings; of course, in components such as licking there are many repeats of basic movements in a bout (e.g., tongue movements). For present purposes they are treated as a single unit.

As noted above, licking has been subdivided into long (more than $\frac{1}{3}$ second) and short (less than $\frac{1}{3}$ second) bout lengths. The duration of licking affects prediction of subsequent behavior. For example, short licking is much more likely to be followed by overhands than by circling or single strokes, whereas long licking is followed almost equally by overhands, circling, and single strokes. The influence of licking duration on subsequent overhands versus circling is more clearly demonstrated in Table 1 which includes data from additional mice and subdivides licking into three durations: short (less than $\frac{1}{3}$ second), medium ($\frac{1}{3}$ to $\frac{2}{3}$ second), and long (more than $\frac{2}{3}$ second).

The interesting point made by these data on licking is that transitions between components of behavior may be partially a function of the duration of these components. Put another way, information about duration may increase predictability of transitions. The development of sophisticated methods for dealing simultaneously with both duration and sequential coupling would be of obvious interest for both the behavioral and neural sciences (cf. Cane, 1961; Chatfield and Lemon, 1970). Elegant approaches to this general problem are currently being pursued in a study of sequential coupling of eye movements in the African chameleon by Mr. J. Mates, a graduate student in our laboratory.

The data presented in Figure 10 are an overall statement of temporal coupling between face-grooming components in the entire face-grooming sequence. It is usually assumed that the probability of particular transitions is constant for the entire sequence (see above). It would be of value to find higher order groupings ("units") of face

Table 1. THE INFLUENCE OF LICKING DURATION ON SUBSEQUENT BEHAVIOR

Duration of licking bout	Behavior immediately following	
	Overhand	*Circle*
Short (N = 73) (Less than $\frac{1}{3}$ sec.)	47 (64%)	9 (12%)
Medium (N = 43) ($\frac{1}{3}$ - $\frac{2}{3}$ seconds)	23 (53%)	11 (26%)
Long (N = 24) (More than $\frac{2}{3}$ sec.)	6 (25%)	7 (29%)

grooming that include the components described in which the probability and perhaps sequential linkage of individual components is partially a function of the higher order units in which they are found.
Thanks to the very able assistance of Miss F. Stilwell, we now seem to
be in a position to suggest not only that such higher order "units"
exist, but that they follow one another in a relatively regular sequence

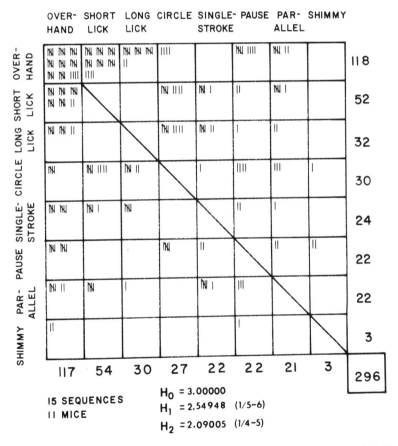

FACE GROOMING COMPONENTS
C 57 CONTROLS

FIGURE 10. Sequential linkage of face-grooming components in grouped C57
control mice. Preceding behaviors are represented by rows, subsequent behaviors by columns.

during the course of face grooming. (It should be emphasized that these "units" represent a tentative grouping and are at present moderately heterogeneous. With few exceptions, however, they can be readily distinguished from one another, and clusters of grooming elements can be assigned to one or another grouping without ambiguity.)

Grooming sequences very often begin with a series of circling and shimmying movements with interjected pauses. We have called this unit 1. Next, shimmys drop out and circling becomes associated with periods of licking and overhead movements. Many of the licking bouts are long at this stage (greater than $\frac{1}{3}$ second, indicated by a dot after the L in Fig. 11). We call this unit 2. The long licking and overhands at this stage may ocassionally appear without circling and still be classified as unit 2 (e.g., lines 1 and 5). Pauses may also occur (e.g., line 6), but are now uncommon, and one gets the impression that grooming is becoming intense. At the end of period 2 and during the start of period 3, for example, one can actually move a mouse with a pencil with only a momentary, if any, disruption of grooming. If the animal stops grooming at all, it will start again almost immediately and in a manner indistinguishable from the grooming shown prior to the interruption. Prior to the end of period 2, however, grooming is easily disrupted and may not start again for a long time. The same is true for later sections of grooming.

A period of single stroking typically occurs next, sometimes associated with parallel strokes and short licking. This is typically the first occurrence of either parallel strokes or single strokes. Long licking is never observed. We call this unit 3. This is usually followed by a period consisting of either overhands separated by short licking or consecutive overhands. Long licking is absent, as are circles, shimmys, single strokes, pauses, and parallels. This is also an intense phase of grooming, and we have called it unit 4. After unit 4 the regularity and often vigor of movements begins to decline. Parallel strokes usually occur at this stage, often mixed with overhands, licking (either long or short), and pauses. Because the internal structure of this unit is less tight than in those occurring previously, we altered our sequence terminology and labeled it 00.

We have found that both very early in grooming and at the termination of a grooming sequence it is difficult to classify movement components into reliable units. However, note that this unpredictability of organizational detail comes at predictable times. Early and late grooming are less structured than grooming during the intervening period. Grooming at the beginning of the sequence (and sometimes at the end; see line 4) contains a mixed cluster of pauses, overhands, and sometimes short licking bouts. Because we are not entirely happy about our

GROOMING UNITS

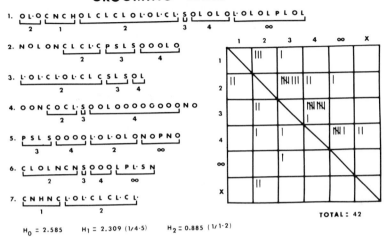

$H_0 = 2.585$ $H_1 = 2.309 \; (1/4\text{-}5)$ $H_2 = 0.885 \; (1/1\text{-}2)$

FIGURE 11. Arrangement of face-grooming components into higher order groupings ("units"). Raw data from entire face-grooming sequences in C57 control mice given in lines 1-7. C = circle; H = shimmy; L = short lick ($< \frac{1}{3}$ sec.); L° = long lick ($> \frac{1}{3}$ sec.); N = pause; O = overhand; P = parallel; S = single stroke (cf. Fig. 9). Four major units represented by numbers 1-4. Less organized units represented by X and 00. Sequential relation between units given in graph at right. Information measures on higher order units given in bottom line.

attempts to classify these components to date, except by exclusion from the other categories, we have lumped them together under the label X. Note that parallels do not occur in X as they do in 00.

This covers all parts of a face-grooming sequence. Some actual examples are presented in Figure 11 in addition to the statistical summary. Notice that the value of H2 is less than 1, which strongly supports the suggestion that these units follow one another in a relatively regular manner (a higher order grouping still). There is also a slight tendency to revert back to a previous unit (oscillate) in a manner similar to that discussed earlier (e.g., Fig. 1) and occasionally to skip a unit in sequence (but rarely more than one).

The units and their coupling might be compared to the phrase structure in human language, which in turn partially determines the order in which individual words will appear (cf. Lenneberg, 1967; Vowles, 1970). Similar principles can probably be found in a variety of behaviors, and they are an important supplement to sequential analyses confined only to elements such as presented in Figure 10. Put another way, one's ability to predict the animal's next behavior when told the current behavior is much enhanced if one is also told the phrase or unit in which that behavior occurs. For example, circling is typically followed by pause or shimmy in unit 1, but by either licking or overhand in unit 2; overhands are followed by overhands only in unit 4.

When these higher order categories are referred to as "units," the term should not be taken too strictly, for there not only can be some variations in the internal organization of these units, but also the boundary between successive units may be graded, rather like the bands of our "rainbow" noted previously. For example, in the first item shown (line 1), the "boundary" of unit 4 would have been extended to the first parallel stroke if the first licking in 00 had been short rather than long. However, this degree of variability is probably a major feature of behavior in higher organisms, and attempts to form categories that are too rigid can prevent rather than facilitate sophisticated analysis (cf. Hinde, 1970). The units described above can be reliably recognized by different observers, and the temporal transitions between these units are quite regular. The distinction between "component" and "unit" as used here is primarily a distinction of degree rather than kind. For example licking is obviously made up of many subcomponents and might itself be treated as a unit. Also repeated occurrences of the same "component" are not necessarily identical in form, and thus the component itself is a category with some internal heterogeneity. For instance, successive overhands in a face-grooming sequence tend to be longer in both duration and excursion; they are not identical. The heuristic value of treating different levels of organizations is unchanged by these considerations, however. Specific cases will be given below.

The apparent (but superficial) paradox is this. One must formulate categories as an aid to analysis, but in complex behavioral systems the categories can rarely be treated as either unitary (indivisible) or entirely and sharply separable from one another.

Central-Peripheral Interactions in Grooming Behavior

Considerable space has been devoted to descriptions of the context and integrative control of grooming behavior because such descriptions are important to the precise analysis of mechanisms. In the next sections of this paper, experiments relevant to central versus peripheral control of grooming will be reviewed briefly. The previous discussion suggests that this may not be an absolute and unitary dichotomy, and that subtle and dynamic interplay between central and peripheral mechanisms might be expected.

When one asks, "Why does a mouse groom?", one obvious answer is that it itches. Indeed, several recent authors have attempted to explain occurrences of grooming in animals as response to peripheral irritants (e.g., Rowell, 1961). However, an alternative model is that the occurrence and sequential ordering of grooming is largely under central control. Simple evidence for the latter model can be obtained by placing peripheral irritants on the animal. In a series of experiments we have found that mild irritants (e.g., light air stream, water drop) placed on the back of a mouse can actually increase *face* grooming. This is largely because in the process of turning to groom the back the mouse initiates its grooming in the normal way—with its face. It is true that the ratio of body grooming to face grooming increases, but both the increase in total face grooming and the normal face . . . back grooming sequence suggest that central sequencing plays a role. With a strong irritant (e.g., ether) on the back, however, the face grooming may be omitted from the initial sequence (the animal goes directly to its back). In this latter case the animal's response becomes more specifically focused on the source of irritation. Obviously peripheral irritants are relevant.

Trigeminal nerve lesions

T. Jones and the author have recently obtained more direct and conclusive information on this problem. It is possible to denervate all or part of the face of the mouse by sectioning appropriate sensory branches of the trigeminal nerve as they course out to their terminations on the face. The effectiveness of the lesions can be readily confirmed by probing the face of the recovered mouse with a needle. Surgery is performed under anesthesia with visual control. A more detailed report of this work is in preparation, but our first set of experiments (Fig. 12) can be summarized briefly. Data from DBA and C57 mice [five of each strain per group—L (trigeminal lesion), S (sham operated), C (control)] are combined for convenience since the major findings were parallel for the two strains. It is apparent that lesioned animals, who had lost sensitivity on their faces, continue to face groom. We were surprised to find that face grooming in the home cage was even increased for some period of time by the lesion. This appears to be partially a response to surgery *per se*, for sham-operated animals (lesions not affecting the face) also showed some increase in face grooming after the operation.

We have also found that body grooming can be increased by both trigeminal lesions and sham operations. This is probably because se-

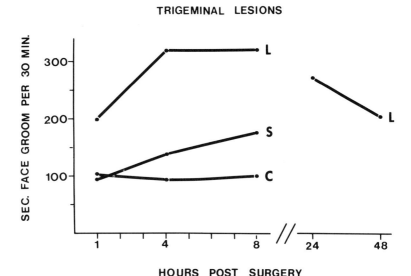

FIGURE 12. Effects of lesions of sensory branches of trigeminal nerve upon face grooming in combined C57 and DBA mice during 30-minute test trials. L = lesioned group; S = sham operated animals; C = control animals.

quential linkage between face and body grooming is in large part centrally controlled. As shown in the right hand side of Figure 12, face grooming in the lesioned animals declines toward the control level by 48 hours after the surgery. We have confirmed this trend in subsequent experiments, although some increase in total face grooming may remain for a period of weeks under certain circumstances.

Two additional observations are relevant here. First, lesioned animals are more likely than controls or shams to show isolated single face swipes in addition to prolonged grooming bouts. These single swipes were included in our analysis, and they account for some of the increase observed. Thus the overall pattern of grooming can be modified. The second observation is most interesting. It is when the animals are in their home cages and undisturbed that the major differences between lesioned and nonlesioned individuals are observed. However, we found it totally impossible to distinguish between lesioned and nonlesioned animals when they were tested in a moderately "stressful" situation, such as a small plastic box not much larger than the animals themselves. Such conditions produce much vigorous grooming (see above) with very stereotyped and prolonged sequences. One cannot escape the impression that the animal is agitated to do something, and not having any obvious course of directed action, grooms. (It is rather like watching people fidget in a dentist's office.) Ethologists have called this type of "apparently irrelevant" behavior "displacement activity," although one must exert caution in the interpretation of mechanisms (see above and Hinde, 1970). What is of interest for our purposes is that the relative importance of changed peripheral cues appears to be minimized under these circumstances.

These latter data suggest that the balance between central and peripheral mechanisms in the control of behavior can shift with the organism's internal state, and when the organism is at a high, but as yet unspecified, motivational setting the degree to which peripheral information sources are processed can be reduced. Under these circumstances, well-established central programs appear to predominate. Again, while comparisons across species and situations must be made with caution, analogous information has been found in the study of human performance (e.g., Broadbent, 1958; Posner, 1966), and this information may have interesting implications for integrative functioning of the central nervous system in higher organisms. It is known, for example, that central activity, even in "motor" systems, can modify neuronal responses to sensory information (e.g., Jasper, 1963; Adkins et al., 1966). It is possible that the mechanisms here are functionally similar to the "narrowing of focus" discussed in a previous section of this paper. Dynamic interactions between mechanisms are again apparent.

Dorsal root lesions

In addition to sensory information originating externally, the central nervous system of mammals receives feedback information from the muscles via the dorsal roots in the spinal cord. What happens if we remove all or part of this information by sectioning the dorsal roots? Dr. M. Rosdolvsky and the author made a series of unilateral lesions from C4 or 5 to T4 in DBA and C57 mice. These lesions remove tactile input and proprioceptive input from the limb from the shoulder down. We have also made a series of less extensive bilateral and unilateral lesions (C6-T2) in these mice, and included a group of sham controls with identical surgery except that the dorsal roots were preserved. Again, probes with a needle of the lesioned mice confirmed complete loss of peripheral sensitivity in the appropriate areas. These experiments are being extended at the present time, but the initial results are clear.

The most striking result is that grooming appears to be basically unchanged in the majority of animals as recorded through normal observation methods. In the few animals where grooming was disrupted, we have independent evidence of spinal cord damage (e.g., hind legs, which should be unaffected by the operation, show residual paralysis). These results suggest that endogenous mechanisms are largely responsible for the patterned output, as has been seen in a variety of invertebrate preparations (e.g., Maynard, 1955; Hoyle, 1964) and for mammals perhaps most convincingly in the complex yet highly stereotyped movement patterns of deglutition (Doty and Bosma, 1956; Doty, Richmond, and Storey, 1967). Other data compatible with a central control model have been obtained for well-established and stereotyped movement sequences, such as hind leg scratching in the rat and cat (Gorska and Jankowska, 1959) and song in various bird species (Konishi, 1965; Nottebohm, 1967).

While overall grooming sequences in our mice appeared to be relatively unaffected by dorsal root lesions, other behaviors clearly were affected. For example, when normal mice swim in a small tank they use their forepaws when approaching the sides; dorsal root sectioned mice do not. Normal mice often explore the sides of a cage by rearing on their hind legs and "patting" the sides with their forepaws; lesioned mice hold their deafferented limb down to one side. When held by the tail, normal mice reach out and grasp onto small objects with both limbs; lesioned mice hold the deafferented limb back and only reach out with it after the normal limb makes contact. Also they do not grasp effectively with the deafferented limb. Even locomotion over rough terrain appears to be more affected in lesioned mice than is grooming. Thus it is obvious that deafferentation can have a differential effect upon different behaviors. One suggestion is that behaviors requiring novel orientation are more dependent upon feedback than well-established stereotyped sequences (cf. Keele, 1968; Konorski, 1967, 1970). A complementary suggestion is that the movements most severely affected involve visual-motor coordination (Hinde, personal communication).

One must obviously be extremely careful about saying that a lesion has no effect upon a given behavior, for this may indicate insufficiently precise observational techniques. Therefore, an extensive series of film analyses (taken at 64 f.p.s.) was made with the assistance of Miss F. Stilwell. A sample summary of face-grooming components in unilaterally lesioned C57 mice is shown in Figure 13. When compared to Figure 10 for normal C57 mice, it is apparent that all the normal grooming components are present. The symmetry in Figure 13 (indicating alternation between two behaviors) is even more marked than in Figure 10, and largely as a result the H2 value is somewhat reduced (i.e., average uncertainty about the next behavior is less with knowl-

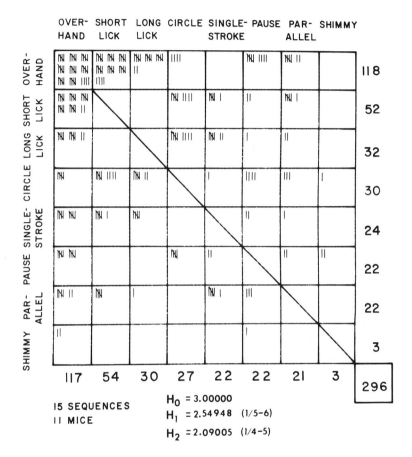

FIGURE 13. Face-grooming components and their sequential relations in C57 mice with dorsal root lesions (cf. Fig. 10).

edge of the previous behavior). While differences with this many degrees of freedom and sample size must be interpreted with caution, it is interesting to note that our deafferented animals consistently showed fewer repeated overhands with reference to the total number of grooming strokes than did the controls (Mann-Whitney $U = 21$; $p < .02$). This cautions that there may be subtle differences produced by the lesion.

Overhands are particularly interesting in this context since they consist of relatively large amplitude and long duration movements, and thus they might be expected to be particularly susceptible to modification by derangement of proprioceptive and tactile feedback. The table in Figure 14 demonstrates that the mean duration of overhand strokes in the deafferented mice is nearly twice that of the controls. For briefer movements, such as single strokes, this difference was not found. This suggests the possibility that one role of proprioceptive feedback in the overhands might be a tonic acceleration of movement. Such has previously been reported both for locust flight (Wilson, 1961, 1966) and swimming movements in the dogfish (Roberts, 1969). This stresses that proprioception should not be assumed to have a single (phasic) function. The variance of time spent in overhand movements in the deafferented mice is also much higher than for the controls (Figure 14). Interestingly, the intact limb in unilaterally deafferented mice shows similar changes in duration and variance of overhand strokes in comparison to control animals, which strongly suggests a remaining coupling between the two limbs. The story is not yet complete, however.

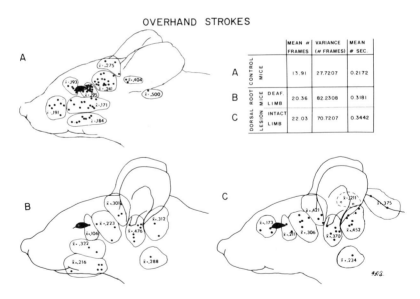

FIGURE 14.   Placement of overhand strokes in grouped control C57 mice (A), the deafferented limb of dorsal root sectioned C57 mice (B), and the intact limb of the same unilaterally lesioned C57 mice (C). Dots represent points of furthest backward excursion of individual overhand movements. Mean times of overhand strokes (seconds) given within boundaries drawn for different limb positions.

For example, as seen in the drawings in Figure 14, the placement of the forelimb at the extent of its backward excursion is less regular in the lesioned animals than in controls (A = control mice; B = deafferented limb, and C = intact limb in lesioned mice), and in lesioned animals there is a tendency to concentrate movements toward the back of the head. The furthest backward excursion is shown by the dots on the drawings. Notice also that the duration of an overhand movement appears to be partially a function of the distance traveled (which would be anticipated if velocity were constant). This is indicated by the mean duration drawn for each cluster of dots. Thus, part of the increase in overhand duration may be due to an increased tendency to make longer strokes; at present the two measures cannot be separated.

Other modifications in the overhand strokes can be seen in Figure 15. The numbers indicate the frame number at which the overhand stroke reached its point of furthest backward excursion and the frame number at which the stroke terminated. The path taken by the limb is drawn. Notice that in comparison to controls (upper row) the deafferented limb may travel a course with considerable irregularity (lack of smooth outline) and may also momentarily halt, or become stuck, during its descent to the nose. This latter point is represented by the dot marked 26-31 in the middle figure of the bottom row, meaning that the limb stopped moving for six frames (approximately 1/10 second). Again, it is of interest that the intact forelimb in the unilaterally deafferented mouse may show similar modification (right hand drawing); some type of cross coupling between the limbs is apparent.

One of our most interesting findings is that although deafferented mice show all of the major elements of grooming (Figure 13), we were unable to determine reliable higher order categories as we had done with the normal mice (Fig. 11). Thus it is possible that these higher order units, and their sequencing, are particularly susceptible to modification of normal feedback pathways.

It has recently been suggested that even complex movements in primates can be established and performed without any proprioceptive input (Taub and Berman, 1968), although at present some caution in interpretation appears called for (e.g., Bossom and Ommaya, 1968; Evarts, 1971). This possibility is interesting because it provides a significant alteration of previous views of integrated movement control in vertebrates (e.g., Mott and Sherrington, 1895). It has become increasingly apparent that some levels of motor programming must contribute even to a variety of rather complex human skills (e.g., Keele, 1968). The point of the above experiments is that central-peripheral interactions are multiplicative and rather subtle, and it is thus necessary to make multiple and precise behavioral measures on different classes of movement prior to the establishment of any sweeping generalizations.

Limb lesions in infant mice

It would be of interest to make similar lesions in infant mice, for there is the possibility that peripheral and proprioceptive stimuli are important for the establishment of a movement sequence, such as grooming, but that once the behavior has become fixed the sequence of movements becomes more autonomous from these sources of stimulation. It is well known that certain species of birds will not develop normal songs without auditory information, but that once these songs are established subsequent deafferentation may have little effect (Konishi, 1965; Marler and Hamilton, 1966; Nottebohm, 1967). At present our technique of infant lesions of sensory pathways in mice has not been sufficient to make precise conclusions. However, a more simple experiment has yielded some interesting insights.

It is possible to amputate the forelegs of infant mice with a simple cut. If all of the mice in a litter are handled and given an application of ferric subsulfate (used to prevent bleeding in the amputated animals), the mother will readily accept and care for the entire litter when it is returned. We have performed a series of such operations in DBA and C57 mice shortly after they are born (within 12 hours). These mice will obviously not be able to perform normal face grooming. It is therefore possible to find out whether basic movement components of face grooming will develop and persist. Laboratory mice show rudimentary face grooming, usually in the form of single swipes at the nose, by their second or third day. Rather complete face grooming can be seen by day 10, and by day 14 the behavior is common.

Figure 16 shows a C57 mouse 14 days of age which had its two forelimbs amputated at the elbow approximately four hours after birth. The animal is performing a series of face-grooming movements even though the amputated limbs cannot reach the face. A most striking characteristic of this mouse, and all others that have had similar surgery to date, is the coordination between the limbs and the tongue. Normal mice lick by bringing the forepaws back to the mouth from a position in front of the face, and protruding the tongue at the appropriate time. This mouse could nearly reach the mouth with its elbows. However, film records have demonstrated that it is not until the elbows are pulled back to a position that would be appropriate for licking in a normal mouse that the tongue typically protrudes from the mouth. The tongue is therefore extended into the air; it does not make contact with the limbs (Fig. 16). Coordination between the upper arms and the tongue movements appears normal, but as a consequence effective licking cannot take place. Other grooming components also occur in the characteristic manner, although there may be some minor alterations that deserve further analysis. It was particularly surprising to discover that the delimbed mice often precede and/or follow grooming by actual licking of the cage floor, or sides, or even another mouse. In contrast to the previous linkage between tongue and limb movements, these last mentioned occurrences of licking appear to be elicited when

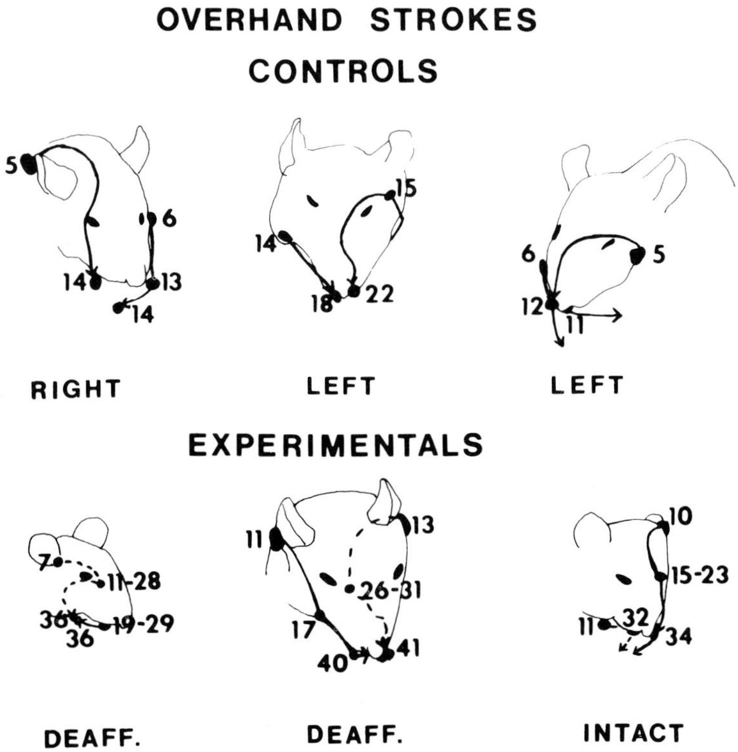

OVERHAND STROKES
CONTROLS

RIGHT   LEFT   LEFT

EXPERIMENTALS

DEAFF.   DEAFF.   INTACT

Figure 15.   Paths of representative overhand strokes in control and experimental (dorsal root sectioned) C57 mice. Numbers represent frame count of motion picture film (64 f.p.s.) when limb was in position indicated by dot. The deafferented limb in the experimental animals followed a course indicated by dashes; the pathway followed by intact limbs is indicated by solid lines. If a limb became stuck in one position for a period of time, this is indicated by the pairs of numbers next to the dot representing position (e.g., 11-28 in the experimental mouse on the left). Note the comparative irregularity of movement in the experimental mice.

the forelimbs are stationary. Thus the degree of (central) linkage between movements can differ under different conditions. Results also suggest that the absence of sensory feedback on the tongue from the limbs may result in the animal seeking this feedback by licking other objects. This again demonstrates that the balance between central and peripheral mechanisms is indeed subtle, and that satisfactory evaluation can be made only by breaking behavior into component parts.

Results of these experiments also indicate that much of the central linkage between movements can be established without normal experience subsequent to birth. Again, however, this does not justify any sweeping conclusion that grooming movements are all genetically preprogramed in any simple sense. For instance, in addition to the elicitation of special licking sequences (above), lesioned animals develop a tighter, more angular sitting posture than normal mice. This means that animals amputated at the elbows can indeed perform rudimentary face washing with the intact upper arm. Simple nature-nurture dichotomies are similar in kind and just as misleading as the dichotomous separation of endogenous and exogenous control processes (cf. Hinde, 1970; Lehrman, 1970; Marler and Hamilton, 1966). In recent years there has been increasing evidence that many basic neuronal mechanisms are highly specified by an organism's genetic makeup (e.g., Weiss, 1950; Sperry, 1965; Jacobson, 1967). However, there is other striking evidence for plasticity, such as observations reported by Prechtl (1965) that even rudimentary human limb reflexes can be modulated by position of the infant *prior to* birth. Behavior in a phenotype, and both genetic and environmental information sources are critical to its development.

Aspects of Movement Pattern Ontogeny

Broad surveys of behavioral development in the mouse have been provided by Williams and Scott (1953) and Fox (1965). Report of our own observations will be restricted to aspects of the emergence and functional coupling of movement sequences related to material in previous sections of this paper. As mentioned above, grooming is one of the first behaviors to develop during ontogeny. Early face-grooming strokes resemble nursing movements and may be developmentally related to them. Face-grooming strokes seen at three to six days usually occur in isolation, are of relatively long duration in comparison to adults, and are difficult to categorize with the criteria used for adults. For example, when mice are approximately five days of age they show many face strokes that are somewhat intermediate between overhands and single strokes. Early single strokes occupy approximately one-fifth

FIGURE 16. Grooming movements in a 14-day-old C57 mouse that had the upper front limbs amputated shortly after birth. Note extension of tongue which was coupled with arm movements.

second as opposed to one-tenth second for adults. Other movements are slower to a similar degree. By ten days of age the strokes are essentially adult in form (although still slower) and are strung together in characteristic sequences (although total face grooming still consists of relatively short bouts). Body and back grooming develop later, and are seen in a relatively adult form by 14 to 16 days. This is largely due to the animals' difficulty in maintaining balance prior to this time.

One gets the impression that many movement components are less tightly coupled in young mice than in adults. When mice are between four and seven days of age, rhythmic movements in a single limb can be elicited by tapping that limb with a blunt probe. It is as if that single limb attempts to run away while the others remain stationary. In older animals the limbs are much more tightly coordinated with one another. Similarly, if a four- to seven-day mouse is placed on its back, three limbs may flail about while one (usually a hind limb) remains stationary.

T. Templeman, an undergraduate student, has performed a very nice descriptive analysis of the ontogeny of swimming behavior in C57 mice that clearly illustrates fundamental changes in movement control. Mice are able to swim during their first day. Initial swimming movements involve all four legs, but the coupling between legs is very loose throughout the first week. Very often one leg, normally a rear leg, will remain stationary, while the other three legs go through swimming movements. Forelimbs are initially more smoothly coordinated than hindlimbs. The swimming movements in the hindlegs increase in frequency from one every two-thirds second within 24 hours after birth, to one every one-half second by day 8, to one every one-third second by day 12, to one every one-seventh second by day 22. The tightness of coupling between limbs (i.e., predictability of movements in all limbs by observation of a given limb) increases steadily during the first ten days of life. Most interestingly, the use of the forelegs when the animal is far from the sides of the tank drops out between 10 and 14 days, and the animal is propelled by the hindlegs only. The forelegs are used primarily to reach for the sides of the tank in adult mice. These changes in coordination are not dependent upon prior practice, although it appears that practice in swimming can accelerate the rate of change. In summary, these observations indicate that first the four limbs become more tightly coupled (one is not "left behind"—all move synchronously), but then the forelegs subsequently become disengaged from the total movement pattern. Thus we may see *both* increased coupling of individual movements and increased differentiation of movements during ontogeny.

Observations on face grooming support this conclusion. In early grooming the forepaws may not reach the tongue during the licking phase, thus again suggesting some early endogenous coupling of limb and tongue movements. By the time the mouse is nine days of age, licking has become effective, although various basic strokes are still often a) slow, b) of short trajectory, and c) isolated (i.e., not sequentially combined with other grooming elements). At this age the animal may also show a series of face-washing movements with only one forepaw; the other may be held in a stationary position on the floor. This is never seen in adult animals. Thus the two forelegs may operate more independently of one another in young mice than in older animals. On the other hand, we have often observed that some young mice combine scratching movements with the hind limb with face washing of the ipsilateral forelimb. This combination is never seen in adult animals.

There has been considerable discussion in the literature as to whether major changes in behavioral ontogeny involve progressive differentiation of movements from a "total pattern" involving most if not all of the body's musculature (e.g., Coghill, 1929) or the progressive combination or "synthesis" of individual patterns that occur early (e.g., Carmichael, 1934; Windle, 1944). Part of the difficulty may be due to the fact that different animals (Coghill's work was with amphibia; Carmichael and Windle worked with mammals) and sometimes different types of movements were investigated. The above observations suggest that the dichotomy may not be useful even for given movements of a given species, for there seems to be both systematic coupling and uncoupling of movements during the course of ontogeny. To date there have been no detailed analyses of the *sequential* coupling of movement patterns during ontogeny, although with the laboratory assistance of C. Ide preliminary evidence has been obtained that similar principles may apply. Working from a standardized battery of behavioral tests, Ide is currently undertaking computer analyses of these developmental stochastic parameters.

When a mouse approximately 48 hours old is removed from its litter and placed on a table for observation, it will use the forelimbs in a rather inadequate way to pivot in one direction or the other. As the paw which is on the outside of the circle moves toward the face, it may make several face-grooming movements before pivoting is reinitiated. One gets the impression that this face brushing is largely triggered by the proximity of the paw to the face, as if a sudden shift in programing occurs. Similarly, if a mouse is placed on its back up until the age of four days, it will flail the four limbs out of the side of the body in a rather uncoordinated fashion, and then suddenly bring a forepaw to the face and groom. Here the proximity of the limb to the face seems to be a less adequate explanation. Rather, it appears that a variety of motor activity can result in sudden and surprising shifts to grooming, which at that stage is a predominant behavior.

Anokhin (1961) has suggested that at different stages in ontogeny different behaviors become predominant, and at that time can be elicited by a variety of different stimuli. In our laboratory we have seen, for example, that a given stimulus can elicit hind-leg scratching, face washing, or general treading movements partially as a function of the mouse's age. Thus at different stages of ontogeny, coupling between different inputs and outputs may shift. A variety of stimuli will often facilitate a newly predominant behavior with subsequent differentiation.

The predominance of face grooming over time in individual DBA mice of the same litter is illustrated in Figure 17. The mice were individually observed in a standard plastic cage for successive ten-minute

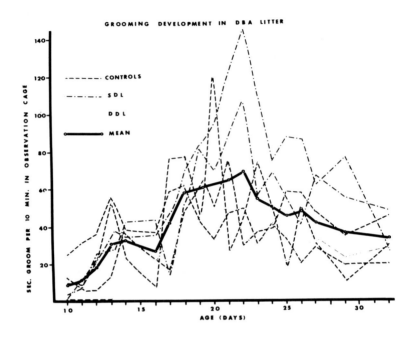

FIGURE 17. Development of grooming in DBA litter. SDL = single forelimb removed; DDL = both forelimbs removed. Heavy line = mean duration score for all six animals. Note increase and subsequent decrease in grooming with age.

trials as indicated. Two mice in this litter had one forelimb removed (SDL) and one mouse had both forelimbs removed (DDL), as discussed above. Notice that the mice show a peak in their grooming between day 18 and day 22. This peak coincides with a general peak of reactivity in the mice, sometimes referred to as the "popcorn" stage. Mice between day 18 and day 25 show an exaggerated startle response to novel stimuli, which slowly declines thereafter. Peak grooming coincides with this period, and this correlation provides independent support for the postulate that emotional reactivity can lead to an increase in the amount of time spent in grooming behavior. Individual animals that demonstrated the earliest peaks in their overall reactivity also showed the earliest peaks in grooming. The DDL mouse showed least overall grooming in this litter, but subsequent data suggest that this is not always the case (occasionally DDL's show the most grooming). Reasons for this variability are being explored.

Central Mechanisms and Grooming Sequences

With the above data as background, analysis of central mechanisms that contribute to grooming patterns presents a problem. One region of the brain that is likely to participate in movement sequences of this type is the extrapyramidal motor system; specifically, the basal ganglia (e.g., Jung and Hassler, 1960). Using this as an initial guide, Mrs. N. Mohler and the author have found it possible to elicit face-grooming sequences by electrical stimulation of the globus pallidus. Stimulation of the same region can also elicit other stereotyped movements, such as chewing, partially as a function of the parameters of stimulation and partially as a function of the animal's behavioral set (cf. von Holst and von St. Paul, 1963; Valenstein, 1969). We have found that different mice can give different results, but that many of these differences can be predicted by prior observation of the mouse's behavior. This is particularly true if the mouse is observed as it goes under anesthesia. For example, some mice do much face grooming as they go under, while others show a much higher level of chewing. It is as if the different behaviors have differential predominance in different animals. Mice that chew while going under anesthesia also tend to chew when stimulated, while mice that spend a greater proportion of their time grooming during anesthesia are more likely to groom to the electrical stimulation.

Obviously, electrical stimulation studies must be interpreted with caution since highly abnormal and synchronized patterns of neural activity are produced. Indeed one might argue that the behaviors which are elicited survive this disruption better than others. This in itself is not entirely without interest since behavioral tests suggest that grooming might be facilitated when other, more complex, behaviors are blocked. However, there is a degree of localization within the nervous system of the effects produced, and when interpreted with caution electrical stimulation techniques can provide useful insights.

Inbred mice provide a unique potential for analysis of central mechanisms underlying patterned behavior that we have recently begun to exploit; numerous mutant strains exist with genetically produced abnormalities in selected central structures. L. Northup in our laboratory has begun to apply previous analyses of face grooming and related behaviors to mice of the strain "Nervous" (Sidman and Green, 1970). Careful genetic and histological analyses by Sidman and Green have demonstrated that Nervous mice are produced by an autosomal recessive which produces atrophy of 90 percent of the Purkinje cells in the lateral hemispheres of the cerebellum and 50 percent of the Purkinje cells in the vermis. Purkinje cells can be thought of as the output path-

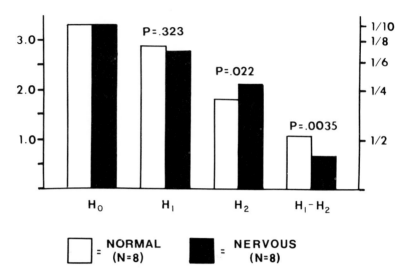

ORGANIZATION OF
FACE GROOMING COMPONENTS

FIGURE 18. Relative probability (H1) and sequential coupling (H2) of ten face-grooming components in "Nervous" and control BALB mice. Information measures indicate significantly less redundancy in temporal patterning of face-grooming components in "Nervous" mice (Purkinje cell atrophy).

ways of the cerebellum. It has long been recognized that the cerebellum in general and the Purkinje cells in particular may play an important role in the temporal integration of behavior (e.g., Eccles, et al., 1967; Evarts and Thach, 1969; Freeman, 1969; Konorski, 1970), but the behavioral measures themselves have not kept up with the sophistication of data at the electrophysiological level.

One of our most interesting findings to date emphasizes the need for careful behavioral description. Nervous mice show all the normal face-grooming components displayed by their normal BALB littermates, and indeed may at first appear indistinguishable from the control animals. However, as illustrated in Figure 18, Nervous mice show a significantly more random sequential linkage between these components than do normal mice. In short this suggests that the Purkinje cells of the cerebellum may contribute not only to the control of individual movement patterns but also to the way these individual components are strung together over time. In further support of this view, Northup has found that while wetting and circling are reliable predic-

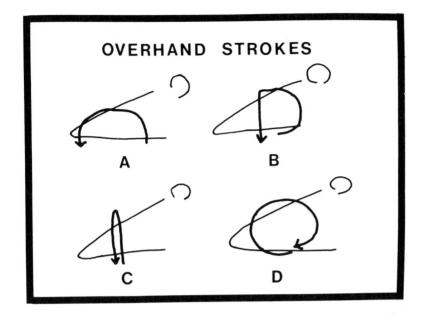

OVERHAND STROKES

A   B

C   D

FIGURE 19.   Pathway of normal overhand stroke (A) and three irregularities (B, C, D) most commonly observed in "Nervous" mice.

tors of subsequent major grooming movements (overhand, single stroke, parallel stroke) in normal mice, they are not reliable predictors in Nervous mice. Nervous mice also stay in higher order grooming patterns for less time (see above), although they may enter these patterns more frequently and the temporal coupling of these higher order units is less regular. Long-term changes in the patterning of other behaviors during the course of a test trial are also less pronounced in Nervous mice than in littermate controls. In a very real sense Nervous mice are more ergodic.

Patterning of individual movements can also be abnormal in Nervous mice. The probability of abnormal movements is partially a function of the duration and extent of the movement, as might be expected. A sample single-frame film analysis of abnormal overhands in Nervous mice showed that 8/22 (or more than one third) of the movements defined as long were abnormal, whereas 53/271 (or less than one fifth) of the movements defined as short were abnormal. The abnormalities are sketched in Figure 19, and they may involve a sudden vertical descent (B), vertical ascent and descent (C), or even movement in the opposite direction to normal (D).

It is obvious that just as there is a hierarchy of levels of analysis that must be considered in behavioral biology, so is there a hierarchy of time periods that must be recognized and integrated. There is the level of muscle contraction, for example, in which we find occasional ataxia in the Nervous mice. There is the level of movement path which, as in overhands, can be affected by cerebellar disorders. There is the problem of linkage between individual elements, the problem of organization of higher order temporal "units" (see above), and the problem of the temporal linkage of these higher order units. Then there is the problem of the sequential linkage of grooming of the face with grooming of the belly, of all grooming and other behaviors (e.g., locomotion). When this time scale is extended further, one gets into problems of shifts in motivational state and then into development itself. In these latter cases problems of information storage become important supplements to problems of integrative functioning. What these components of behavior are, how they fit together in a hierarchial pattern, and how they are read out in adaptive temporal patterns are all related questions of considerable importance. Detailed analyses of movement sequences in inbred and hybrid mice should provide some useful insights.

ACKNOWLEDGEMENTS: This paper is dedicated to Professor Robert A. Hinde, who for many years has served as teacher, colleague, and friend, and whose many contributions inspired much of the work reported herein.

The research reported herein was supported in part by funds from the Office of Scientific and Scholarly Research, University of Oregon Graduate School, and in part by Public Health Service grants MH-16887 and MH-16955. Undergraduate research program participants T. Jones, N. Mohler, and T. Templeman received support from National Science Foundation funds awarded to the Department of Biology, the University of Oregon.

My thanks to Miss F. Stilwell for her assistance in data collection and preparation of figures, Dr. F. Delcomyn for assistance with the Linc 8 program, and H. Howard for printing the illustrations.

Literature Cited

Adkins, R. J., R. W. Morse, and A. L. Towe. 1966. Control of somatosensory input by cerebral cortex. Science, 153: 1020-1022.

Altman, S. G. 1965. Sociobiology of Rhesus monkeys. II. Stochastics of social communication. J. Theor. Biol., 8: 490-522.

Anokhin, P. K. 1961. Contributions to discussion: inborn and reflex behaviour. In Brain Mechanisms and Learning. J. F. Delafresnaye, ed., pp. 643-644. Oxford: Blackwell.

Attneave, F. 1959. Applications of Information Theory to Psychology. New York: Holt.

Bossom, J., and A. K. Ommaya. 1968. Visuo-motor adaptation (to prismatic transformation of the retinal image) in monkeys with bilateral dorsal rhizotomy. Brain, *91:* 161-172.

Broadbent, D. E. 1958. *Perception and Communication.* London: Pergamon.

Cane, V. 1961. Some ways of describing behavior. In *Current Problems in Animal Behaviour,* W. H. Thorpe and O. L. Zangwill, eds., pp. 361-388. Cambridge: Cambridge University Press.

Carmichael, L. 1934. An experimental study in the prenatal guinea-pig of the origin and development of reflexes and patterns of behaviour in relation to the stimulation of specific receptor areas during the period of active fetal life. Genet. Psychol. Monogr., *16:* 337-491.

Chatfield, C., and R. E. Lemon. 1970. Analyzing sequences of behavioural events. J. Theor. Biol., *29:* 427-445.

Coghill, G. E. 1929. *Anatomy and the Problem of Behavior.* Cambridge: Cambridge University Press. New York: Macmillan.

Darwin, C. 1872. *The Expression of the Emotions in Man and Animals.* London: John Murray.

Doty, R. W., and J. F. Bosma. 1956. An electromyographic analysis of reflex deglutition. J. Neurophysiol., *19:* 44-60.

Doty, R. W., W. H. Richmond, and A. T. Storey. 1967. Effect of medullary lesions on coordination of deglutition. Exp. Neurol., *17:* 91-106.

Eccles, J. C., M. Ito, and J. Szentagothai. 1967. *The Cerebellum as a Neuronal Machine.* Berlin-Heidelberg-New York: Springer-Verlag.

Evarts, E. V. 1971. Central control of movement. Neurosciences Res. Prog. Bull. 9.

Evarts, E. V., and W. T. Thach. 1969. Motor mechanisms of the CNS: cerebrocerebellar interrelations. Ann. Rev. Physiol., *31:* 451-498.

Fentress, J. C. 1965. *Aspects of Arousal and Control in the Behaviour of Voles.* Ph.D. thesis, Cambridge University.

Fentress, J. C. 1967. Observations on the behavioral development of a hand-reared male timber wolf. Am. Zoologist, *7:* 339-351.

Fentress, J. C. 1968a. Interrupted ongoing behaviour in two species of vole (*Microtus agrestis* and *Clethrionomys britannicus*). I. Response as a function of preceding activity and the context of an apparently 'irrelevant' motor pattern. Anim. Behav., *16:* 135-153.

Fentress, J. C. 1968b. Interrupted ongoing behaviour in two species of vole (*Microtus agrestis* and *Clethrionomys britannicus*). II. Extended analysis of motivational variables underlying fleeing and grooming behaviour. Anim. Behav., *16:* 154-167.

Fox, M. W. 1965. Reflex-ontogeny and behavioural development of the mouse. Anim. Behav., *13:* 234-241.

Freeman, J. A. 1969. The cerebellum as a timing device: an experimental study in the frog. In *Neurobiology of Cerebellar Evolution and Development,* R. Llinas, ed., pp. 397-420. Chicago: American Medical Association Education and Research Foundation.

Garner, W. 1962. *Uncertainty and Structure as Psychological Concepts.* New York: Wiley.

Gorska, T., and E. Jankowska. 1959. Instrumental conditioned reflexes of the deafferented limb in cats and rats. Bull. acad. poloni. Sci., *7:* 161-164.

Guttman, R., Lieblich, I., and Naftall, G., 1969. Variation in activity scores and sequences in two inbred mouse strains, their hybrids, and backcrosses. Anim. Behav., *17:* 374-385.

Hinde, R. A. 1969. Control of movement patterns in animals. Q. J. Exp. Psychol., *21:* 105-126.

Hinde, R. A. 1970. *Animal Behaviour: A Synthesis of Ethology and Comparative Psychology.* New York: McGraw-Hill.

Hinde, R. A., and J. G. Stevenson. 1969. Sequences of behavior. In *Advances in the Study of Behavior,* Vol. 2, D. S. Lehrman, R. A. Hinde, and E. Shaw, eds., pp. 267-296. New York: Academic Press.

Hinde, R. A., and J. G. Stevenson. 1970. Goals and response control. In *Development and Evolution of Behavior,* L. R. Aronson, E. Tobach, D. S. Lehrman, and J. S. Rosenblatt, eds., pp. 216-237. San Francisco: W. H. Freeman.

Hinde, R. A., and N. Tinbergen. 1958. The comparative study of species-specific behavior. In *Behavior and Evolution,* A. Roe and G. G. Simpson, eds., pp. 251-268. New Haven: Yale University Press.

Holst, E. von, and U. von Saint Paul. 1963. On the functional organization of drives. Anim. Behav., *11:* 1-20.

Hoyle, G. 1964. Exploration of neuronal mechanisms underlying behavior in insects. In *Neural Theory and Modeling,* R. F. Reiss, ed., pp. 346-476. Stanford: Stanford University Press.

Jacobson, M. 1967. Starting points for research in the ontogeny of behavior. In *Major Problems in Developmental Biology,* M. Locke, ed., pp. 339-383. New York: Academic Press.

Jasper, H. H. 1963. Studies of non-specific effects upon electrical responses in sensory systems. In *Progress in Brain Research,* Vol. 1: *Brain Mechanisms.* G. Moruzzi, A. Fessard, and H. H. Jasper, eds., pp. 272-293. Amsterdam: Elsevier.

Jung, R., and R. Hassler. 1960. The extrapyramidal motor system. In *Handbook of Physiology,* section 1, *Neurophysiology,* vol. 2, J. Field, H. W. Magoun, and V. E. Hall, eds., pp. 863-928. Washington, D.C.: American Physiological Society.

Keele, S. W. 1968. Movement control in skilled motor performance. Psychol. Bull., *70:* 387-403.

Konishi, M. 1965. The role of auditory feedback in the control of vocalization in the white-crowned sparrow. Z. Tierpsychol., *22:* 770-783.

Konorski, J. 1967. *Integrative Activity of the Brain.* Chicago: University of Chicago Press.

Konorski, J. 1970. The problem of the peripheral control of skilled movements. Intern. J .Neuroscience, *1:* 39-50.

Lashley, K. S. 1951. The problem of serial order in behavior. In *Cerebral Mechanisms in Behavior,* L. A. Jeffress, ed., pp. 112-136. New York: Wiley.

Lehrman, D. S. 1970. Semantic and conceptual issues in the nature-nuture problem. In *Development and Evolution of Behavior,* L. R. Aronson, E. Tobach, D. S. Lehrman, and J. S. Rosenblatt, eds., pp. 17-52. San Francisco: W. H. Freeman.

Lenneberg, E. H. 1967. *Biological Foundations of Language.* New York: Wiley.

Luria, A. R. 1966. *Higher Cortical Functions in Man.* New York: Basic Books.

Marler, P., and W. S. Hamilton. 1966. *Mechanisms of Animal Behavior.* New York: Wiley.

Maynard, D. M. 1955. Activity in a crustacean ganglion. II. Pattern and interaction in burst formation. Biol. Bull., *109:* 420-436.

Mott, F. W., and C. S. Sherrington. 1895. Experiments upon the influence of sensory nerves upon movement and nutrition of the limbs. Preliminary communication, Proc. Roy. Soc., *57:* 481-488.

Nelson, K. 1964. The temporal patterning of courtship behavior in the glandu-locaudine fishes (*Ostariophysi, Characidae*). Behaviour, *24:* 90-146.

Nottebohm, F. 1967. The role of sensory feedback in the development of avian vocalizations. Proc. Int. Ornithol. Congr., *14:* 265-280.

Posner, M. I. 1966. Components of skilled performance. Science, *152:* 1712-1718.

Prechtl, H. F. R. 1965. Problems of behavioral studies in the newborn infant. In *Advances in the Study of Behavior,* Vol. 1, D. S. Lehrman, R. A. Hinde, and E. Shaw, eds., pp. 75-99. New York: Academic Press.

Roberts, B. L. 1969. Spontaneous rhythms in the motoneurons of spinal dogfish (*Scyliorhinus canicula*). J. Mar. Biol. Assn. (U.K.), *49:* 33-49.

Rowell, C. H. F. 1961. Displacement grooming in the chaffinch. Anim. Behav., *9:* 38-63.

Sevenster, P. 1961. A causal analysis of displacement activity, fanning in *Gasterosteus aculeatus.* Behaviour, Suppl. 9.

Sidman, R. L., and M. C. Green. 1970. "Nervous," a new mutant mouse with cerebellar disease. Symp. of the Centre National de la Recherche Scientific, Orleans-la-Source, pp. 49-61.

Sperry, R. W. 1952. Neurology and the mind-brain problem. Amer. Scientist, *40:* 291-312.

Sperry, R. W. 1965. Embryogenesis of behavioral nerve nets. In *Organogenesis,* R. L. De Haan and H. Ursprung, eds., pp. 161-186. New York: Holt.

Taub, E., and A. J. Berman. 1968. Movement and learning in the absence of sensory feedback. In *The Neuropsychology of Spatially Oriented Behavior,* S. J. Freedman, ed., pp. 173-192. Homewood: Dorsey Press.

Valenstein, E. S. 1969. Biology of drives. In *Neurosciences Research Symposium Summaries,* Vol. 3, F. O. Schmitt, T. Melnechuk, G. C. Quarton, and G. Adelman, eds., pp. 1-108. Cambridge: The Massachusetts Institute of Technology.

Vowles, D. M. 1970. Neuroethology, evolution, and grammar. In *Development and Evolution of Behavior,* L. R. Aronson, E. Tobach, D. S. Lehrman, and J. S. Rosenblatt, eds., pp. 194-215. San Francisco: W. H. Freeman.

Weiss, P. 1950. Experimental analysis of coordination by the disarrangement of central-peripheral relations. *Symp. Soc. Exp. Biol., 4:* 92-111.

Williams, E., and J. P. Scott. 1953. The development of social behavior patterns in the mouse, in relation to natural periods. Behaviour, *6:* 35-65.

Wilson, D. M. 1961. The central nervous control of flight in a locust. J. Exp. Biol., *38:* 471-490.

Wilson, D. M. 1966. Central nervous mechanisms for the generation of rhythmic behaviour in arthropods. *Symp. Soc. Exp. Biol., 20:* 199-228.

Wilz, K. J. 1970. The disinhibition interpretation of the "displacement" activities during courtship in the three-spined stickleback, *Gasterosteus aculeatus.* Anim. Behav., *18:* 682-687.

Windle, W. F. 1944. Genesis of somatic motor function in mammalian embryos: a synthesizing article. Physiol. Zool., *17:* 247-260.

Genetic and Behavioral Studies of *Drosophila* Neurological Mutants

WILLIAM D. KAPLAN
Division of Biology, City of Hope
National Medical Center
Duarte, California

JUST AS THE SEQUENTIAL STEPS in a biochemical pathway have been dissected one by one through the study of single gene mutations, it may be possible to unravel complex behavior by altering the individual elements one at a time by means of specific mutations. In the field of the behavioral or neurological genetics of *Drosophila*, the primary interest has, until recently, been placed upon a multifactorial approach. The multifactorial or biometrical approach has successfully exploited the use of selection experiments to demonstrate that such behaviors as phototaxis, geotaxis, and spontaneous activity have a large heritability component (Ewing, 1961; Connolly, 1966; Dohzhansky and Spassky, 1969), and measurements have been made on the contributions of individual chromosomes (Hirsch, 1967). Ultimately, however, it becomes difficult to unravel the genetics of these high- and low-response lines, since their differences depend upon the recombination, and additive effects, of many genes.

Mating behavior has been studied extensively in *Drosophila*, and the considerable literature has been reviewed most recently by Manning (1964). But here again one is faced by a complex situation, since sensory and motor systems as well as interactions between two individuals are involved. Furthermore, studies that have utilized single gene changes and their effects upon mating behavior have shown that these effects have resulted from anatomical or activity changes resulting from pleiotropism in a trivial, psychologically unimportant sense (Wilcock, 1969).

Benzer (1967) designed an elegantly simple technique for producing sex-linked single gene phototactic mutants. Through subsequent studies utilizing neurophysiological and genetic techniques, he uncovered five cistrons on the X chromosome of *Drosophila* associated with various abnormalities in visual function (Hotta and Benzer, 1969, 1970). All the phenotypes proved to be autonomous in the genetic sense. In mosaic individuals a mutant eye functions abnormally, regardless of the amount of normal tissue present elsewhere, indicating that the primary lesion, responsible for the behavioral defects of the mutant flies, lies within the eye.

We have been studying several sex-linked mutants known as "shakers" because of their characteristic reaction to diethyl ether. During the period of anesthetization, the legs of the flies shake in a regular, rhythmic fashion. Neurophysiological studies with one of these mutants, *Hyperkinetic^{1P}* (*Hk^{1P}*) have shown that the leg-shaking action is governed by rhythmic bursts of motor nerve impulses (Ikeda and Kaplan, 1970a). This patterned motor nerve activity is endogenous within the thoracic ganglion because neither the cephalic ganglion, the neuromuscular system, nor the sensory input to the thoracic ganglion plays a role in the pattern formation. The active areas furnishing this patterned activity were localized within three pairs of motor regions in the thoracic ganglion. However, the detection of a local abnormality does not necessarily locate the primary lesion, since the mutant properties of the motor neurons may represent a secondary result of malfunction elsewhere. The result may be autonomous within the motor neurons themselves, or it may reflect the lack (or presence) of a circulating metabolite. This may be tested by creating composite individuals by the technique of genetic mosaicism. Gynandromorphs, mosaic for male tissue carrying the mutant gene hemizygously and female tissue which is heterozygous for the gene in questions, make it possible to investigate this matter. Studies on gynandromorphs, mosaic for *Hk^{1P}*, have shown that the expression of the gene is autonomous, unaffected by any substance circulating in the body fluid or by the genotype of other parts of the fly (Ikeda and Kaplan, 1970b).

The question of genetic background is important in studying the effects of a single gene change. As far as it is possible, one must assure that the genetic backgrounds of the mutant and control stocks are identical so that observed behavioral differences may be attributed to the action of the mutant gene in question, and not to modifiers or background differences. To this effect the shaker stocks have been created in such a way that they are as similar in background to the *Canton-S* reference stock in which the mutants were produced as it is possible, by existing techniques, to achieve (Kaplan and Trout, 1969; Ikeda and Kaplan, 1970a). However, two identical stocks maintained side by side will in time diverge from each other by the accumulation of spontaneous mutations. Thus, periodically, the mutant stock is crossed back to the *Canton-S* stock and new lines of each are reestablished.

Descriptions of Shakers

The shakers to be described in this report were found in an experiment in which ethyl methane sulfonate was the mutagenic agent. The experiment was designed for the detection of maternally influenced lethals (Kaplan, Seecof, Trout, and Pasternack, 1970). At a point in the procedure it was necessary to examine the contents of individual vials, at which time the shakers were detected. Four mutants, derived from four different treated males were uncovered. One was localized to the originally described shaker locus (Catsch, 1944) and was designated *Shaker⁵*. The others represent new loci as follows:

Hyperkinetic¹ᴾ (Hk¹ᴾ)	*30.9*
Hyperkinetic²ᵀ (Hk²ᵀ)	*30.4*
Ether à go go (Eag)	*50.0*
Shaker⁵ (Sh⁵)	*58.2*

Shaking appears to be a semi-dominant character, since *Shaker-Canton* hybrids shake less markedly than homozygous shakers.

Activity level of shakers

While handling cultures of shakers, it was observed that they were more active than *Canton S* populations. There is more movement within culture bottles, and the populations require a longer settling time after the cultures have been disturbed. Activity levels have been measured by placing a vial with flies near a microphone, isolated from external light and sound. The number of buzzes made by the flies may then be monitored through earphones or recorded electronically by a counter. Under these conditions it is found that shaker flies, unetherized, produce more buzzes over a period of time than the reference *Canton-S* wild-type stock (Kaplan and Trout, 1969). *Hk¹ᴾ* and *Sh⁵* are the two most active stocks; *Hk²ᵀ* and *Eag* have about half the activity as defined by this measure. Shakers also react to the presence of each other. A single fly produces a level of activity of about two buzzes per minute at a given age. Two flies produce four buzzes per minute per fly. Thereafter, the level of activity remains the same per fly with increasing numbers, until the recording device fails to distinguish between one or two or more buzzes occurring simultaneously, at which point activity appears to fall (Kaplan and Trout, 1968).

Quantitative studies on leg shaking

Data derived from Hk^{1P}, Hk^{2T}, Sh^5, and combinations of these specific mutants will be presented. *Ether à go go* poses special problems and will not be reviewed here.

The shaking described here as a response to ether is not a waking phenomenon. Flies kept in a closed system with a defined concentration of ether will reach a steady state and will continue to shake for as long as 24 hours, until they die of desiccation. On the other hand, shakers as well as normal flies when entering and emerging from ether anesthetization make vigorous struggling movements as they attempt to right themselves. These movements are quite different from shaking.

Females were used in these studies because comparisons were made between the individual mutants and several combinations requiring two X chromosomes. In *Drosophila* an individual with two X chromosomes is, by definition, female.

Measurements were done on the right midleg of individual flies. Each female was etherized, then pasted on her back to a stiff card; all legs but the right midleg were pasted down; the fly was then given thirty minutes to recover from the anesthesia. At the end of this period, the card was placed in a test tube stoppered with a hollow polyethylene stopper, and ether was injected into the tube to give the desired ether vapor, v/v, concentration of 3 percent.

At first, there are rapid, struggling movements, but by forty minutes, when measurements are started, the shaking has reached a steady state which persists for hours. For some experiments the end segment of the leg (tarsus) was monitored (Figure 1). In others the femur was fixed in position and the movement of the tibia alone was measured (Figure 2). The first procedure represents the summation of the movements of all leg segments.

Quantitative data are derived in two ways. In both cases, the test tubes containing the flies are mounted on the stage of a compound microscope and the image of the moving leg is focused upon a ground-glass screen. Movements are monitored by a photo-electric cell which triggers a pulse generator. Each back-and-forth movement of the leg across the photo cell produces a pulse. By the first method, the pulse generator is connected to a pulse-averaging device similar to a count rate meter. The output of the pulse-averaging device is sufficiently am-

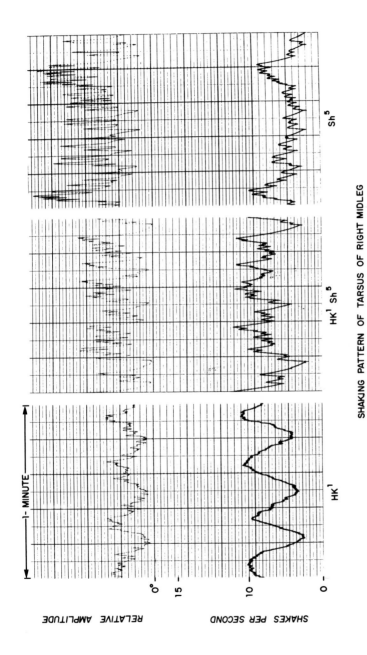

SHAKING PATTERN OF TARSUS OF RIGHT MIDLEG

FIGURE 1.   Time course of shaking, right midleg, lower tracing, averaged frequency; upper tracing, averaged amplitude, no absolute vertical scale. Flies were five days old.

plified to drive a two-channel recording potentiometer. One channel records the frequency of leg movements, and the other records the amplitude of the movements, which represents the excursion of the leg across the exposed area of the photo cell (Figure 1). In order to calibrate the measurements, a fly leg was fixed to a speaker cone which was then driven by a sine square wave generator over a range of determined frequencies.

The second method, which records greater detail, makes use of an FM tape recorder, to which the pulse generator is connected. The tape recording is subsequently played back, at a slower speed, to an event recorder (Figure 2).

In addition to Hk^{1P}/Hk^{1P}, Hk^{2T}/Hk^{2T}, and Sh^5/Sh^5 homozygotes, we have studied females which represent combinations of the several mutants. Also, we have monitored the leg shaking of flies which carried Hk^{1P} or Hk^{2T} in one X chromosome and a deficiency $(Df(1)v^{74})$, which covers the hyperkinetic locus, in the other X chromosome. These deficiency heterozygotes carry only one mutant hyperkinetic gene, in contrast to the Hk^{1P}/Hk^{1P} homozygotes, which carry two mutant genes per genome. By use of the deficiency, it is possible to obtain information concerning the action of the mutant gene. Since no gene is present in the deficiency, no hyperkinetic gene product is contributed by the deficiency chromosome. If Hk^{1P} and Hk^{2T} were amorphs, mutants producing no gene product (Muller, 1932), Hk^{1P}/Df should be equivalent to Hk^{1P}/Hk^{1P}, and Hk^{2T}/Hk^{2T} should have the same phenotype as Hk^{2T}/Df. Any differences between the respective phenotypes, however, would indicate the mutant gene is coding for a gene product, either one different from the normal gene (neomorph) or less of the normal product (hypomorph).

Because the Hk^{1P} locus is about 27 crossover units from the Sh^5 locus, gene recombinants are easily obtained, yielding a chromosome carrying both mutant genes, $Hk^{1P}Sh^5$. By crossing the double mutant to Hk^{1P}/Hk^{1P} or Sh^5/Sh^5, the number of Hk^{1P} and Sh^5 genes relative to each other may be varied, yielding a range of genotypes to test the influence of one mutant upon the other.

Figures 1 and 2 show the shaking patterns for Hk^{1P}, Sh^5, and $Hk^{1P}Sh^5$, Figure 1 by the first recording method described, Figure 2, by the second. Each spike of Figure 2 represents one back-and-forth leg movement.

From these records three measurements may be obtained:

a) The number of shakes per minute; b) the number of shaking bursts or episodes per minute; and c) the percent of time spent shaking.

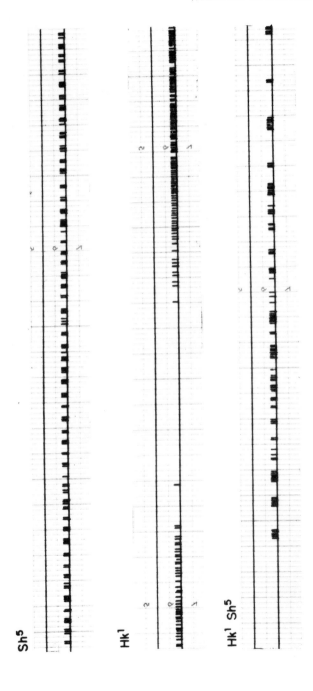

FIGURE 2. Time course of shaking, right mid-leg. Each shake represents one back-and-forth movement. Flies were five days old.

For the hyperkinetic locus, these parameters are highly correlated
($r > 0.9$). The genotype with the greatest number of shakes per
minute is active the greatest portion of time, and has fewer, but longer,
bursts per minute. The most active flies do not simply shake faster,
they shake for longer periods of time and consequently have fewer
quiet periods over a given interval of time. These data are summarized
in Table 1 and graphed in Figure 3.

It may be seen that the patterns of shaking are different for the
several mutants. Hk^{1P} has a steady cycling pattern; Hk^{2T} is like Hk^{1P}
except that there are shorter bursts of shaking. Sh^5 has a fairly steady
rate, shaking with short bursts superimposed over the basic pattern,
and it is active virtually 100 percent of the time. This is reflected in
the appearance of a characteristic jerkiness when Sh^5 is viewed under
the microscope. The double mutant exhibits a pattern of shaking made
up of components of both Sh^5 and Hk^{1P}. The Hk^{1P} cycling pattern is
present with the little Sh^5 wavelets superimposed (Figure 1). More-
over, the silent periods of Hk^{1P} are present, indicating that Hk^{1P} must
be inhibiting Sh^5 (Figure 2). Therefore, the mutant genes appear to
be codominant, but assuming the absence of shaking to be the normal
condition, $Hk^{1P}Sh^5$ is more normal than either mutant by itself
(Table 1).

The combinations of genes at the hyperkinetic locus may be
ranked using the percent of time spent shaking as a measure of mutant effect.

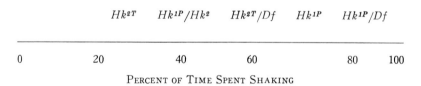

Hk^{2T}	Hk^{1P}/Hk^2	Hk^{2T}/Df	Hk^{1P}	Hk^{1P}/Df
0	20	40	60	80	100

PERCENT OF TIME SPENT SHAKING

Hybrids of Hk^{1P} and *Canton*, or Hk^{1P} and other shakers, show
very mild shaking. Hk^{1P}/Hk^{2T}, however, exhibits active shaking, more
than Hk^{2T} homozygotes. These two genes do not complement each
other in this and other activity measurements and are thus most surely
allelic.

The ranking of genotypes according to the percent of time spent
shaking shows that the mutant genes Hk^{1P} and Hk^{2T} are not amorphs,
since their respective heterozygotes with the deficiency have a greater
abnormality than either homozygote; that is, two doses of either mutant gene result in an expression closer to normal than does a single
dose.

Table 1. Shaking Pattern of Hk^{1P} and Hk^{2T} and Their Combinations with Sh^5; 3 Percent Ether Vapors Monitored After Forty Minutes Exposure

	Hk^{1P}				Hk^{2T}		
	Shakes per min.	Bursts per min.	Percent time active		Shakes per min.	Bursts per min.	Percent time active
$\frac{Hk^{1P}+}{Hk^{1P}+}$	418	3.4	65	$\frac{Hk^{2T}+}{Hk^{2T}+}$	227	6.0	27
$\frac{Hk^{1P}+}{Df(1)v^{74}}$	521	2.1	83	$\frac{Hk^{2T}+}{Df(1)v^{74}}$	486	4.3	59
$\frac{Hk^{1P}+}{Hk^{2T}+}$	282	4.6	37	$\frac{Hk^{1P}Sh^5}{Hk^{2T}Sh^5}$	336	122.0	42
$\frac{Hk^{1P}+}{Hk^{1P}Sh^5}$	442	3.4	77	$\frac{Hk^{2T}+}{Hk^{2T}Sh^5}$	342	4.2	33
$\frac{Hk^{1P}Sh^5}{Hk^{1P}Sh^5}$	269	3.9	37	$\frac{Hk^{2T}Sh^5}{Hk^{2T}Sh^5}$	374	166.0	26
$\frac{Hk^{1P}Sh^5}{+Sh^5}$	300	106.0	35	$\frac{Hk^{2T}Sh^5}{+Sh^5}$	386	166.0	74
$\frac{+Sh^5}{+Sh^5}$	313	140.0	ca 100				
$\frac{Hk^{1P}+}{+Sh^5}$	150		20	$\frac{Hk^{2T}+}{+Sh^5}$	86		24
$\frac{Hk^{1P}Sh^5}{++}$	——— Inactive ———			$\frac{Hk^{2T}Sh^5}{++}$	89		5

In summary

The following points may be made to summarize the data on leg-shaking patterns:

(*1*). Hk^{1P} has fewer shaking cycles or episodes than Hk^{2T}: 3/minute vs. 6/minute.

(*2*). The individual episodes are longer for Hk^{1P}. Therefore, Hk^{1P} is actively shaking more of the time: 40 vs 15 seconds within each minute.

(*3*). Sh^5 has continuous short bursts, little cycling as with Hk^{1P} and Hk^{2T}: about 140/minute.

(*4*). Hk^{1P}, Hk^{2T}, and Sh^5 hybrids with *Canton-S* do not shake under the conditions of these tests.

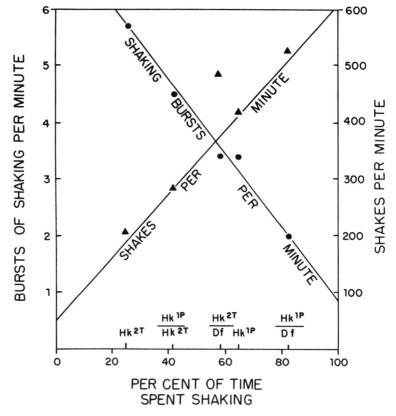

FIGURE 3.   Relation of leg-shaking frequency and frequency of shaking bursts or episodes to the percent of time spent shaking for Hk^{1P} and Hk^{2T}.

(5). With a deletion at the Hk locus, Hk^{1P} and Hk^{2T} have longer cycles, more shakes per minute, and are active more of the time.

(6). $Hk^{1P}+/Hk^{2T}+$ is intermediate between $Hk^{2T}+/Hk^{2T}+$ and $Hk^{1P}+/Hk^{1P}+$; therefore, with respect to shaking, Hk^{1P} and Hk^{2T} are codominant.

(7). $Hk^{2T}Sh^{5}/Hk^{2T}Sh^{5}$ is like $+Sh^{5}/+Sh^{5}$; Sh^{5} is epistatic to Hk^{2T}.

(8). $Hk^{1P}Sh^{5}/Hk^{2T}Sh^{5}$ is similar in phenotype to $Hk^{2T}Sh^{5}/Hk^{2T}Sh^{5}$.

(9). $Hk^{1P}Sh^{5}/Hk^{1P}Sh^{5}$ is a combination midway between $Hk^{1P}+/Hk^{1P}+$ and $+Sh^{5}/+Sh^{5}$.

(10). One dose of Sh^{5} does not affect $Hk^{1P}+/Hk^{1P}+$.

(11). One dose of Sh^{5} makes $Hk^{2T}+/Hk^{2T}+$ more like $+Sh^{5}/+Sh^{5}$.

(12). One dose of Hk^{1P} makes $+Sh^{5}/+Sh^{5}$ more like $Hk^{1P}+/Hk^{1P}+$.

Neurological Basis of Leg-shaking Action

The neural mechanism governing the abnormal response to ether consists of rhythmic bursts of activity produced by motor neurons located within the thoracic ganglion. Extracellular recordings (Figure 4A, upper trace) were used to locate the neurons responsible for the fast rhythmic bursts of the mutant. An extensive search was made over the thoracic ganglion, but active areas furnishing the mutant patern of nerve impulses were limited to very small paired regions in the pro-, meso-, and metathoracic motor areas (Ikeda and Kaplan, 1970).

After the areas of motor activity were located, it was possible to obtain intracellular recordings from single motor neurons. Two types of neurons were encountered. In the presence of ether, type I neurons discharged action potentials without any sign of prepotential. In the case of type II neurons, an action potential was always preceded by a slowly rising depolarization. The type I neurons were encountered about three times more frequently than type II. This observation and physiological considerations (Ikeda and Kaplan, 1970a) indicate a pacemaking role for type II neurons.

One genetic tool available in work with *Drosophila,* which has proved highly informative in physiological studies of Hk^{1P}, is the gynandric or mosaic individual. It is possible by a relatively simple technique to generate gynandromorphs, or flies mosaic for male and female tissue. The crosses may be set up in such a way that the male

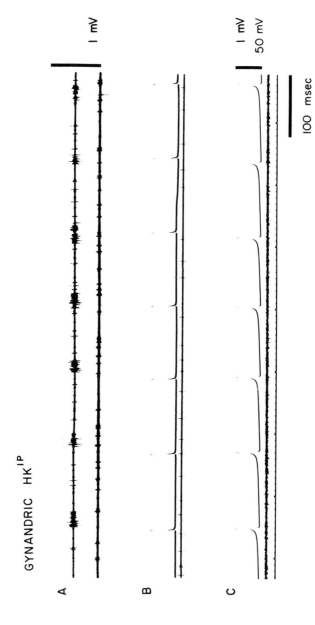

FIGURE 4.   Electrical activities in the thoracic ganglion of gynandromorphs, mosaic for *Hk*[1P]. A. Simultaneous extracellular recordings from paired prothoracic motor regions: upper trace, male side; lower trace, female side. Calibration for A, 1 mV and 100 m sec. B. Simultaneous recordings from a pair of metathoracic motor regions: upper trace, intracellular recording from a type I neuron on the male side; lower trace, extracellular recording from the region on the female side. C. Simultaneous recordings from a pair of mesothoracic regions; upper trace, intracellular recording from a type II neuron on the male side; lower trace, extracellular recording from the region on the female side. (See text for descriptions of gynandromorphs.) Calibration for B and C, 50 mV for upper trace, 1 mV for lower trace; 100 msec.

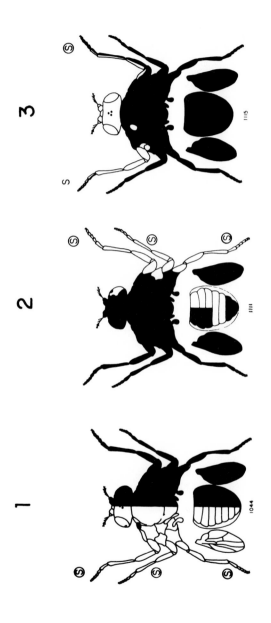

FIGURE 5. Gynandric patterns. (1) Simple bilateral gynandromorph which provided tracings shown in Figure 4(A). (2) Only three right legs and sternopleurite are male. (2) Right prothoracic leg is male on cephalic aspect, female on caudal aspect; left prothoracic leg as well as the head are male. The male tissue is white; female tissue is black. Leg-shaking action observed on legs marked S. Rhythmic bursts recorded from motor regions corresponding to legs with circled S. Recording from the left prothoracic leg of (3) failed for technical reasons.

tissue carries Hk^{1P} hemizygously along with recessive markers for eye and body color and bristle structure, whereas the female carries Hk^{1P} heterozygously. Since the markers are also heterozygous, the female tissue has normal color and conformation. Because of the nature of *Drosophila* embryology, the line establishing the male-female demarcation may vary in position so that a wide variety of mosaic types may be obtained. These mosaics were exploited in carrying out observations on shaking and neurophysiological activity.

Simple bilateral gynandromorphs (Figure 5, 1) provided data as illustrated in Figure 4A. The upper trace is an extracellular recording from the male side of the gynandromorph and the lower trace from the female side. Figure 4B shows, in the upper trace, an intracellular recording from the male side of a mosaic whose head and thorax were divided bilaterally into male and female tissue, but whose abdomen was entirely female. Figure 4C shows in the upper trace an intracellular recording from a type II neuron on the male side; the lower trace is an extracellular recording from the female side of the same mosaic whose head was entirely female, thorax bilaterally divided, and whose abdomen was entirely female. In all of these cases, the male legs (Hk^{1P}), but not the female ($Hk^{1P}/+$), showed shaking activity.

These studies demonstrated clearly that the expression of Hk^{1P} is autonomous in the genetic sense, that the leg movements of the two sides of the fly are governed independently, that the genotype of the head and abdomen have no influence upon the motor areas, and that nothing circulating in the body fluids may mediate the expression of the mutant or wild-type gene, either with respect to the shaking or the firing pattern of the motor neurons.

Figure 5 illustrates mosaics of a more complicated type than the ones described above. In these examples, (2 and 3), the thoraxes are not divided along the midline, so that there may be male legs on an otherwise female thorax or individual legs that are mosaic. Although not figured here, several cases of female legs on an otherwise male thorax have been found. (It is not the maleness or the femaleness of the tissue that is important. Sex is used simply as a marker to distinguish between Hk^{1P} and $Hk^{1P}/+$.)

From a study of these more complicated mosaics, it is possible to conclude that each motor region acts independently of the others. If the cells of any one of the motor regions are Hk^{1P}, the mutant neural activity may be detected and the leg will shake in the presence of ether. Figure 5(2) shows three right legs which are Hk^{1P}, associated with a female thorax. The legs shake and the characteristic Hk^{1P} firing pattern is present in the corresponding regions of the ganglion. The genotype of the thoracic chitin, therefore, has nothing to do with the phe-

nomenon of shaking. Only when the motor region is mutant in genotype will shaking be observed.

Quantitative data with respect to the firing patterns of the motor
neurons agree closely with the data obtained by monitoring the leg-
shaking patterns. Neurons firing from about 180 to 600 times per
minute have been obtained from Hk^{1P} neurons. Bursts of activity have
been recorded lasting from two to ten seconds, which agrees with the
data derived from monitoring the leg movements directly.

Kinetogenic Response of Hyperkinetic Flies

Hk^{1P} and Hk^{2T} flies show an unusual response to movement. When
an object moves above a vial containing these mutant flies, they jump
and fall over. The response may be measured quite simply by the
experimenter moving his hand above a vial containing a single fly and
scoring the number of positive responses in fifty trials. By this measure

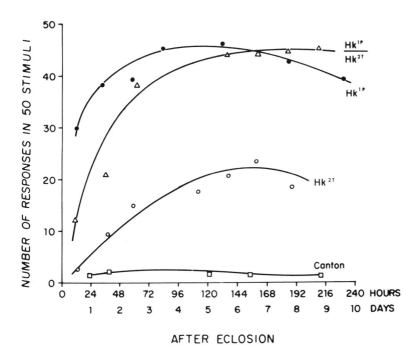

FIGURE 6.   Kinetogenic response in relation to age of Hk^{1P} homozygotes, Hk^{2T}
homozygotes, and Hk^{1P}/Hk^{2T} hybrid.

it is found that Hk^{1P} is more active than Hk^{2T}, and in both cases response increases with age, plateauing at about five days (Figure 6). This response is a recessive component of their behavior since it disappears in hybrids with Canton-S $(Hk^{1P}/+)$ and in compounds with other shakers (Kaplan and Trout, 1969).

On the other hand, Hk^{1P}/Hk^{2T} responds as much as Hk^{1P}/Hk^{1P}, showing that these genes are allelic, since with respect to this behavior, also, there is no complementation and they map to almost identical loci.

Although Sh^5 does not show a consistent kinetogenic response, occasionally an individual fly will jump as much as 20 times out of 50. It was of interest, therefore, to measure the effect of Sh^5 upon Hk^{1P} by testing the various combinations of Sh^5 with these genes. Since the double mutant stock was available, it was possible to create the following genotypes for Hk^{1P} - Sh^5 interactions: $Hk^{1P}+/+Sh^5$, $Hk^{1P}Sh^5/Hk^{1P}+$, $Hk^{1P}Sh^5/Hk^{1P}Sh^5$, $Hk^{1P}Sh^5/+Sh^5$, and $Hk^{1P}Sh^5 / Df +$. Similar ones also were made for Hk^{2T}. Of course, all of these flies were female.

The interaction of Hk^{1P} with Sh^5 is illustrated in Figure 7. Table 2 presents the data from Hk^{1P} and Hk^{2T} interactions with Sh^5.

The simple hybrid, $Hk^{1P}+/+Sh^5$, shows little more than the response of the wild-type Canton-S strain, as is true of any stock in which Hk^{1P} is present heterozygously. However, when Hk^{1P} is in the homozygous condition, the addition of one Sh^5 gene reduces the response. Adding a second Sh^5 gene inhibits the response still further. It is possible to suggest that these observations may be the result of a different set of modifiers brought into the stock when Sh^5 was introduced into the Hk^{1P} chromosome. However, since the same interaction occurs with Hk^{2T} and Sh^5, although starting at a different level, this does not seem likely.

Therefore, to rule out the possibility that a different genetic background is the basis for this phenomenon, we have replaced, by a series of controlled crosses, the chromatin material to the left and right of the Hk^{1P}, Hk^{2T}, and Sh^5 loci with Canton-S chromatin. By this device, the three mutant genes, except for very small regions immediately adjacent to them, are carried in identical Canton-S X chromosomes.

Reference to Table 2 will show, again, that Hk^{2T}/Df and Hk^{2T}/Hk^{2T} are different in phenotype, indicating that the mutant gene must be coding for a particuar gene product. Since Hk^{1P} is already at the upper limit of its response with respect to this behavior, there is no difference between Hk^{1P}/Hk^{1P} and Hk^{1P}/Df. But in all other instances the presence of the deficiency produces a more extreme phenotype

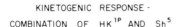

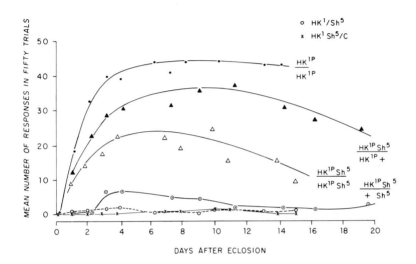

FIGURE 7. Nonallelic interaction in the expression of a behavioral phenotype: kinetogenic response of Hk^{1P} and Sh^5 combinations in relation to age. C indicates the wild-type chromosome of *Canton-S*.

than the Hk^{1P} or Hk^{2T} gene would do in its place. This could mean that the mutant gene is functioning less efficiently than the normal gene, and that two doses of the mutant approach normality to a greater degree than a single one. However, the presence of an altered gene product cannot be ruled out at this point. Unfortunately, the homozygous deficiency is lethal. The lethality is due to the absence of many genes, since the deficiency is a fairly large one.

The nonallelic interaction of Hk^{1P} and Hk^{2T} with Sh^5 to produce a more normal phenotype than either mutant exhibits by itself is characteristic of their expression. This has been seen in activity levels (Kaplan and Trout, 1969), in longevity studies (Trout and Kaplan, 1970), as well as in leg movements and the kinetogenic response. It is interesting that the normal allele of each respective gene does not normalize the expression of the individual mutants as in $Hk^{1P}+/Hk^{1P}+$ and $+Sh^5/+Sh^5$. Only the mutant alleles complement each other.

Table 2.   EFFECT OF Sh^5 GENE ON Hk KINETOGENIC RESPONSE; AVERAGE OF FIFTEEN FLIES (FIVE DAYS) IN EACH DETERMINATION

Sh^5 Genes absent		One Sh^5 gene present		Two Sh^5 genes present	
Genotype	Responses/50	Genotype	Responses/50	Genotype	Responses/50
$\dfrac{Hk^{1P}+}{Df(1)v^{74}}$	41.6 ± 1.7	$\dfrac{Hk^{1P}Sh^5}{Df(1)v^{74}}$	41.9 ± 1.4		
$\dfrac{Hk^{1P}+}{Hk^{1P}+}$	40.9 ± 0.9	$\dfrac{Hk^{1P}Sh^5}{Hk^{1P}+}$	31.0 ± 1.1	$\dfrac{Hk^{1P}Sh^5}{Hk^{1P}Sh^5}$	22.4 ± 1.4
$\dfrac{Hk^{2T}+}{Df(1)v^{74}}$	41.8 ± 1.7	$\dfrac{Hk^{2T}Sh^5}{Df(1)v^{74}}$	41.4 ± 1.4		
$\dfrac{Hk^{1P}+}{Hk^{2T}+}$	43.0 ± 0.7	$\dfrac{Hk^{1P}Sh^5}{Hk^{2T}+}$	31.4 ± 1.7	$\dfrac{Hk^{1P}Sh^5}{Hk^{2T}Sh^5}$	20.4 ± 1.3
$\dfrac{Hk^{2T}+}{Hk^{2T}+}$	26.0 ± 1.2	$\dfrac{Hk^{2T}Sh^5}{Hk^{2T}+}$	21.9 ± 1.6	$\dfrac{Hk^{2T}Sh^5}{Hk^{2T}Sh^5}$	17.0 ± 1.7

Table 3. KINETOGENIC RESPONSES TO (a) MOVEMENT AND (b) FLASHING LIGHT; (a) AND (b) TESTED IN SAME INDIVIDUAL FLIES AGED FIVE DAYS

	$\dfrac{Hk^{1P}}{Hk^{1P}}$		$\dfrac{Hk^{1P}}{Df(1)v^{74}}$		$\dfrac{Hk^{2T}}{Hk^{2T}}$		$\dfrac{Hk^{2T}}{Df(1)v^{74}}$	
	(a)*	(b)*	(a)	(b)	(a)	(b)	(a)	(b)
1	39	24	45	23	21	12	41	10
2	40	12	41	20	24	5	26	3
3	39	20	43	14	26	5	46	13
4	41	13	36	23	22	5	32	5
5	40	23	43	14	11	6	33	4
6	36	12	34	24	25	1	39	2
7	35	23	44	24	34	0	39	5
8	43	24	40	9	27	0	47	19
9	43	17	39	15	19	1	43	8
10	43	20	37	22	18	16	34	3
11	34	22	37	20	16	16	42	3
12	40	24	40	15	16	0	40	22
13	44	16	35	18	25	0	39	9
14	40	17	46	10	26	0	43	7
15			44	20	26	4	35	10
16					21	5		
Mean	39.8	19	40.3	18	22.3	4.8	38.6	8.2
S.e.	0.9	1.3	1.0	1.3	1.4	1.4	1.5	1.5
Percent	79.6	76.0	80.6	72.0	44.6	19.2	77.2	32.8

* (a) 50 trials; (b) 25 trials.

In conducting experiments to measure the kinetogenic response, we have, in effect, a shadow moving above the flies. It is possible, therefore, that the flies respond to a decrease or an increase in light intensity. Turning an incandescent or fluorescent light on and off does not elicit the response. However, a rapidly flashing, extremely bright, strobe light does call forth the kinetogenic response. The response is influenced by both the intensity of the light and the number of cycles per second (Figures 8 and 9). A flash of 100 Hertz is the fastest our equipment can supply and is the most effective stimulus. Below 75 Hertz, the response drops to zero.

Because the duration of the individual flashes remains the same, the interval between flashes increases with decreasing flash frequency. Consequently, over a given period of time, more light is delivered at a higher, as compared to a lower, frequency. Therefore, we cannot be sure at present whether the response occurs only with the high-frequency stimulation or whether it is exclusively an intensity phenomenon.

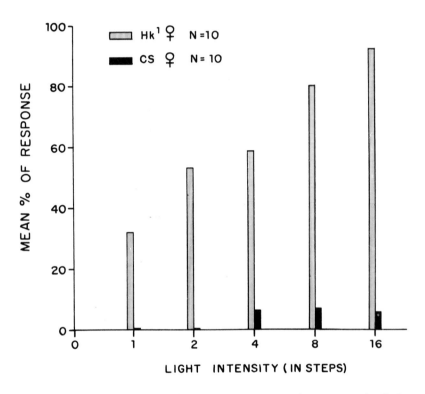

FIGURE 8.   The relation of the kinetogenic response to the intensity of a flashing light. *Canton-S* and *Hyperkinetic* [1P]. Strobe light (Grass Instrument PST2) at two inches from the fly. Intensity scale geometric; approximate maximum flash intensity is 1.5 x 10[6] candle power.

Only Hk^{1P} and Hk^{2T} respond to the flashing light stimulus. Genotypes that are good jumpers in response to movement respond well to flashing light. The correlation between the two effects is good, as may be seen in Table 3, but in the case of individual flies there is some variation. Some flies with a high score in response to movement, jump poorly in response to the flashing light, and some that respond well to

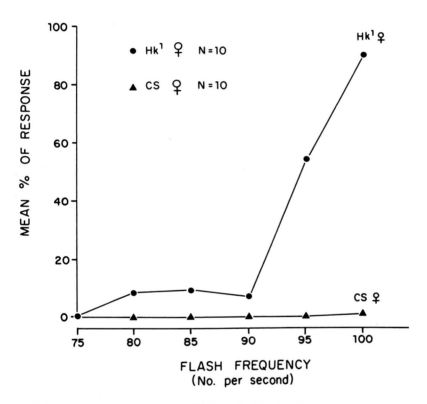

FIGURE 9. The relation of the kinetogenic response of *Canton-S* and *Hyper-kinetic[1P]* flies to the flash frequency. Each stimulus consists of ten flashes at the designated frequency; intensity 8.

the flashing light are poor performers with respect to movement (Table 3).

The kinetogenic response is a complex behavioral pattern requiring, as it does, a motor response to a visual stimulus. Flies carrying *Hk[1P]* and a gene for the character *tan,* which produces blindness, do not exhibit the kinetogenic response. It will be of interest to determine whether the hyperkinetic mutant genes have influenced the neurons in the visual pathway as well as in the motor system. By studying the kinetogenic response of mosaic flies it will be possible to determine whether the visual as well as the motor system must be mutant in order to elicit the response.

Summary

The long-range goal of these studies is ultimately to understand the role of the normal as well as the mutant alleles at the *shaker* gene loci. Perhaps it is too early to construct theories or models with the data presently available except to say that a reasonable approach is to regard the observed behaviors as the manifestation of an imbalance between excitatory and inhibitory neuronal systems.

The neurophysiological work completed establishes that the mutant abnormality is not attributable to a defect in function at the neuromuscular junction. The abnormal electrical activity observed in the ganglion may, possibly, be attributed to a membrane abnormality or to an imbalance between the action of the excitatory and inhibitory systems.

Neurochemical studies on several neurotransmitter systems have been carried out: γ-aminobutyric acid, known to be a true inhibitory transmitter in crustacea and vertebrates and to be present in large quantities in *Drosophila* (Frontali, 1964) ; glutamic acid, believed to be the neuromuscular transmitter in insects (Usherwood, Machili and Leaf, 1968) ; and acetyl choline which occurs at high levels in *Drosophila* (Dewhurst, McCaman and Kaplan, 1970). The levels of these substances have been measured by determining the specific activities of the enzymes concerned with their synthesis and degradation.

In the case of the transmitters tested above, there were no differences in the levels of enzyme-specific activities between the shakers and the *Canton-S* wild-type strain. This tells us that the transmitter substances are produced in normal amounts and destroyed normally, but does not say anything concerning their release from pre-synaptic vesicles or their utilization at post-synaptic sites.

The data derived from study of the mutant combinations indicate that excitatory and inhibitory interactions are occurring. There appears to be a balance between the number of Sh^5 genes and the relative strength of the Hk locus. (Normal $> Hk^{2T}/Hk^{2T} > Hk^{2T}/Df > Hk^{1P}/Hk^{2T} > Hk^{1P}/Hk^{1P} > Hk^{1P}/Df$) Sh^5 genes have the effect of lowering the jumping rate, two Sh^5 genes more than one, but the stocks are still responding in proportion to the mutant potency at the

Hk locus. This relation is true, also, for the shaking patterns. Two Sh^5 genes affect all the Hk combinations studied, but the stronger (most mutant) ones less. One dose of Sh^5 affects the less mutant Hk combinations such as Hk^{2T}/Hk^{2T}, Hk^{1P}/Hk^{2T}, and Hk^{2T}/Df; and has less effect on Hk^{1P}/Hk^{1P} and Hk^{1P}/Df. In many cases flies exhibit both the Sh^5 type of shaking and the Hk type to give a shaking pattern that is the summation of the tyo types (see Figure 1).

One may speculate, therefore, that the product of the wild-type genes, $+^{Hk}$, exerts an inhibitory effect; when it is mutant or missing, the fly can shake in the presence of ether and is more active and responsive to stimuli when awake. Shaking, longevity, jumping, and activity levels are related to the degree of abnormality of the locus.

The same seems to be true for the $+^{Sh5}$ locus. However, since only one allele, Sh^5, is presently available, one cannot say as much about it.

The Hk and Sh^5 loci have normalizing effects upon each other *while mutant*. The more "abnormal" the Sh^5 locus, the more normal is an Hk type fly and vice versa, as regards the phenotypes studied. Since the normal gene of either mutant is not epistatic to the other mutant ($Hk^{1P}+/Hk^{1P}+$ is more mutant than $Hk^{1P}Sh^5/Hk^{1P}Sh^5$), the normal genes do not mutually inhibit each other, only the mutant genes do. Therefore, it would seem that the mutant genes have both inhibitory and excitatory effects.

Biochemical, pharmacological, and neurophysiological studies are presently being done in order to find the ways in which the mutations affect the behavior of the flies, and it is hoped that we can arrive at an understanding of how genetic control is extended to the properties of the individual nerve cell.

ACKNOWLEDGMENTS: It is a pleasure to acknowledge the work of my colleagues who have contributed much to the study of the shaker mutants: Kazuo Ikeda, Ph.D., Eugene Roberts, Ph.D., William E. Trout, Ph.D., and Patrick Wong, Ph.D. The skilled technical assistance of Mrs. A. Colin and Mr. Barry Hanstein is greatly appreciated, as is the typing of Mrs. Shirley Pemberton. The deficiency stock, $Df(1)v^{74}$, was obtained from Professor George Lefevre, San Fernando Valley State College, Northridge, California. The work was supported by grants from the USPHS HS07442 (K.I.) and NB08014 (W.D.K.) and in part by the Harry A. and Alma Kessler Research Fellowship.

Questions and Answers

QUESTION. Is it easier or more difficult to anesthetize shakers as compared with wild-type flies?

ANSWER. They become anesthetized at the same levels of ether as normal flies. However, they may be somewhat more sensitive. They are more sensitive to the killing effects of ether than wild-type flies. They are killed within a shorter period of time with given concentrations than are wild-type flies.

QUESTION. Have you been able to inhibit shaking?

ANSWER. We are getting ready to do this kind of thing. We have played around with it a little but not in a very systematic fashion. The only chemical that has stopped the shaking was ammonia. But then ammonia would probably stop anything so I have tried a few of the antiepileptic agents. The flies were killed in all the concentrations I worked with. We have not tried lower concentrations subsequently. It would be interesting to be able to stop the shaking or to produce shaking in normal flies.

QUESTION. Do you know what the neural basis of the jumping response is?

ANSWER. One interesting approach to answering this question lies in the use of mosaics, to see what relationship of sensory to motor tissue is necessary in order to evoke this response. If the mutation affects the membrane structure of neurons, there is no reason to believe that only motor neurons would be so affected. There may be some influence upon the sensory system that we have not yet been able to detect.

QUESTION. Are there any measurable effects of the mutations upon the larvae?

ANSWER. Dr. Trout devised a very ingenious simple experiment to test the possibility. I think it is a good method but he does not. As you know, when larvae crawl through their world their mouth parts are used to dig in and pull the rest of the body forward. When they are placed upon a soft agar surface sprinkled with charcoal dust, you get a picture of the mouth prints. The difference between the wild type and the mutant was quite striking. But he is not certain the flies were at a comparable physiological stage because the mutants may develop at a different rate from the control stock. But not much has been done on studying the behavior of larval flies. I think people are just beginning to get interested in that.

Literature Cited

Benzer, S. 1967. Behavioral mutants of *Drosophila* isolated by countercurrent distribution. Proc. Nat. Acad. Sci. (U.S.), *58:* 1112-1119.

Connolly, K. 1966. Locomotor activity in *Drosophila.* II. Selection for active and inactive strains. Anim. Behaviour, *14:* 444-449.

Dewhurst, S., R. McCaman, and W. D. Kaplan. 1970. The time course of development of acetylcholinesterase and choline acetyltransferase in *Drosophila melanogaster.* Biochemical Genetics, *4:* 499-508.

Dobzhansky, Th., and B. Spassky. 1969. Artificial and natural selection for two behavioral traits in *Drosophila pseudoobscura.* Proc. Nat. Acad. Sci. (U.S.), *62:* 75-80.

Ewing, A. W. 1963. Attempts to select for spontaneous activity in *Drosophila melanogaster.* Anim. Behavior, *11:* 369-378.

Frontali, N. 1964. Brain glutamic decarboxylase and synthesis of γ-aminobutyric acid in vertebrate and invertebrate species. In *Comparative Neurochemistry,* D. Richter, ed., pp. 185-192. New York: Pergamon Press.

Hirsch, J. 1967. Behavior-genetic analysis at the chromosome level of organization. In *Behavior-Genetic Analysis,* J. Hirsch, ed., pp. 258-269. New York: McGraw-Hill.

Hotta, Y., and S. Benzer. 1969. Abnormal electroretinograms in visual mutants of *Drosophila.* Nature, *222:* 354-356.

Hotta, Y., and S. Benzer. 1970. Genetic dissection of the *Drosophila* nervous system by means of mosaics. Proc. Nat. Acad. Sci. (U.S.), *67:* 1156-1163.

Ikeda, K., and W. D. Kaplan. 1970a. Patterned neural activity of a mutant *Drosophila melanogaster.* Proc. Nat. Acad. Sci. (U.S.), *66:* 765-772.

Ikeda, K., and W. D. Kaplan. 1970b. Unilaterally patterned neural activity of gynandromorphs, mosaic for a neurological mutant of *Drosophila melanogaster.* Proc. Nat. Acad. Sci. (U.S.), *67:* 1480-1487.

Kaplan, W. D., and W. E. Trout. 1968. Activity and reactivity of shaker flies. Genetics, *60:* 191 (abstract).

Kaplan, W. D., and W. E. Trout. 1969. The behavior of four neurological mutants of *Drosophila* Genetics *61:* 399-408.

Kaplan, W. D., R. L. Seecef, W. E. Trout, and M. E. Pasternack. 1970. Production and relative frequency of maternally influenced lethals in *Drosophila melanogaster.* Amer. Nat., *104:* 261-271.

Manning, A. 1965. *Drosophila* and the evolution of behavior. In *Viewpoints in Biology,* J. D. Carthy and C. L. Duddington, eds., pp. 125-169. London: Butterworths.

Muller, H. J. 1932. Further studies on the nature and causes of gene mutations. Proc. 6th Int. Congr. Genet. (1933), *1:* 213-255.

Trout, W. E., and W. D. Kaplan. 1970. A relation between longevity, metabolic rate, and activity in shaker mutants of *Drosophila melanogaster.* Exp. Gerontol., *5:* 83-92.

Usherwood, P. N. R., P. Machili, and G. Leaf. 1968. L-glutamate at insect excitatory nerve-muscle synapses. Nature, *219:* 1169-1172.

Wilcock, J. 1969. Gene action and behavior: an evaluation of major gene pleiotropism. Psychol. Bull., *72:* 1-29.

Lateral Specialization of the Human Brain: Behavioral Manifestations and Possible Evolutionary Basis

Jerre Levy
Department of Psychology
University of Colorado
Boulder, Colorado

THE EVOLUTION OF THE MAMMALIAN BRAIN has been marked by the development of increasingly complex capacities mediated by increasingly complex structures. Yet there has been no essential qualitative leap either in the types of behavior nor in the physio-anatomical functioning of the brain in moving from the rat to the chimpanzee. In contrast, the human brain, unlike that of any other animal, has laterally specialized hemispheres, each hemisphere being proficient in a function for which the other is deficient. In the vast majority of right handers, for example, only the left hemisphere is capable of controlling speech production. Though mammals show a skilled manipulation ability of a preferred hand, it is not clear that this preferred paw or hand usage really reflects a genetically determined cerebral laterization of function.

Studies with several strains of inbred mice reveal that 50 percent of each strain is left-pawed and 50 percent is right-pawed (Collins, 1968). In addition, three generations of selection for pawedness fail to produce a change in these percentages in the F-3 generation (Collins, 1969). Seven generations of selection for pawedness in rats also failed to change the 50-50 pawedness ratio (Peterson, 1934). At least in mice and rats, it seems clear that the determinants of paw usage do not derive from any genetically determined hemispheric dominance, but rather reside in accidental contingencies in the environment which initially reinforce the use of one or the other paw. Studies with monkeys (Ettlinger, 1964) and chimpanzees (Finch, 1941) also show that there are no significant differences in the proportions of left- and right-handed animals, again suggesting a lack of genetic determination.

Handedness in human beings is clearly correlated with hemispheric dominance for speech, and there can be no doubt that there is a large genetic component in the determinants of human hand usage. Various genetic models for handedness have been proposed, none of which is completely satisfactory (Trankell, 1955; Annett 1964, 1967). Nevertheless, it is clear that, at most, a two gene, two allele per gene model is sufficient to account for handedness ratios in various types of matings (Levy and Nagylaki, 1972). Thus, the human brain, unlike other mammalian brains, has laterally specialized hemispheres whose asymmetry results, in large part, from genetic factors. This lateral specialization is most clearly manifested in cases of unilateral cerebral damage which typically produce a loss or decrease of a specific function. Even in the highest anthropoid apes, a unilateral lesion produces no effects except in the contralateral sensory field or contralateral part of the body. No high-level cognitive functions, such as the ability to recognize complex visual stimuli, are lost. In order to produce such a loss in the chimpanzee, bilaterally symmetric lesions must be produced.

Cerebral lateralization therefore represents a rather sudden qualitative leap in evolution. Such leaps are relatively rare events, and one is led to wonder what selective advantage was conferred by functional asymmetry of the hemispheres. Some have suggested that since the vocal apparatus consists of nondivided, unique structures, a single hemisphere would provide less confusing control and that this is the explanation for the fact that language abilities reside in a single hemisphere. This explanation, as appealing as it might seem, suffers from serious deficiencies. In the first place, it is not only language for which hemispheric specialization exists, but certain other functions as well which have nothing to do with motor control. In the second place, there are people who can control speech production from either hemisphere, that is, both half-brains contain language centers, but who, nevertheless, show no stuttering or any other confusion in speech.

The specific nature of right and left hemispheric function in man is not entirely clear. Although the differentiation has been mainly described in terms of language for the left hemisphere (Broca, 1960; Critchley, 1962; Hécaen, 1962) and constructional praxis for the right hemisphere (Paterson and Zangwill, 1944; Piercy and Smyth, 1962; Bogen and Gazzaniga, 1965), these categories leave much to be desired. Both descriptions are in terms of output functions of two "black boxes." Eventually, the real aim is to provide an explanation in anatomical and physiological terms, at present at least in psychological terms, of the mechanisms underlying such output.

Studies with unilaterally brain-damaged patients have provided certain leads as to the possible basis of the output. Paterson and Zangwill (1944) reported on two patients whose right hemispheres had been damaged; the first by a penetrating brain wound, the second from a pony kick. The first patient had great difficulty telling time, and he could only do so by noting the individual positions of the clock hands separately and calculating time. He was normal on verbal intelligence, but was quite deficient on high-grade visual-spatial tasks. He could draw two-dimensional objects, but not complex designs. There was a confusion of perspective, depth, and planes, as well as a disproportion in relative size. He always drew piecemeal, i.e., item by item, and appeared to lack any grasp of the object as a whole. He was quite poor on block design tests. The second patient, like the first, could reproduce two-dimensional shapes, but not complex objects. He was preoccupied with minute details, ignoring the overall configuration. He was deficient on block design tests, and he could identify rooms he had previously seen only by recognition of individual objects. Hécaen and Angelergues (1962), in examining patients in whom the right parietal cortex had been removed for control of epilepsy, found that some of these patients, following surgery, could not recognize faces. Facial agnosia is rarely, if ever, seen following left hemisphere removals.

In the light of findings such as these, it looks as if there is some essential difference between the hemispheres in the methods which are used to process information. It appears that the minor hemisphere is a Gestalt specialist, not particularly interested in the analytic details of the world of sensation, but greatly concerned with general configuration, while, in contrast, the left hemisphere is an expert in symbol translation and analysis, but lacks configurational understanding. The right hemisphere appears to be more highly developed than the left for the type of Gestalt synthesis typically required in perception. It can appreciate spatial configurations and ignore unimportant details. It shows a deficiency, however, in analysis and fails to pay attention to specific focal qualities of stimuli, and it shows a relative inability to deal with symbols. The left hemisphere is not only able to pay attention to detailed features of stimuli, but it can then assign some symbol to represent a given feature. Like a computer, it can analyze and describe the results of its analysis, but, also like a computer, it fails to appreciate the Gestalt. In summary, the two human hemispheres appear to be two specialists—one designed for synthesis, the other for analysis.

The studies which will be discussed here were directed toward confirming or not confirming the preceding ideas as to the nature of hemispheric specialization. The motivating force behind these investigations was the hope that a clearer understanding of the nature of lateral specialization would provide a clue as to the adaptive advantage conferred in evolution by functional differentiation of the human hemispheres.

The subjects[1] were epileptics who had undergone total forebrain commissurotomy, involving section of the corpus callosum, anterior, and hippocampal commissures performed in a single operation. The cerebral hemispheres were thus left with no interhemispheric connections. This surgery, first performed in the 1930's (Van Wagenen and Herren, 1940), had been found to be quite effective in reducing the severity and frequency of epileptic seizures. The present series of patients have likewise benefited from the surgery, and, in some cases, seizures have been abolished altogether.

In such patients the mental capacities of the surgically disconnected hemispheres can be assessed independently by the use of testing procedures that lateralize sensory input, central processing, and/or motor readout to one or the other hemisphere. The separate performance of each of the disconnected hemispheres can then be compared for the same test task. Particularly in patients having a minimum of cerebal damage, the separate testing of each hemisphere of the same individual on the same test performance offers special advantages.

Figure 1 shows a schematic of the human brain after the large bridge of connecting fibers has been cut. Each hemisphere of the brain is mainly associated with one side of the body, the right brain presiding over the left side and the left brain over the right side. All sensations on one side of the body or in one visual half-field are transmitted to the opposite half-brain. Each eye projects to both sides of the brain. The nasal retina of an eye projects to the contralateral hemisphere, and the temporal retina projects to the ipsilateral hemisphere. Any stimulus which appears to the left of a fixation point thus projects to the nasal retina of the left eye and the temporal retina of the right eye, both of

---

[1] These experiments were carried out with a series of patients of Drs. Vogel and Bogen of Los Angeles (1962, 1963; Bogen, Fisher, and Vogel, 1965) in Roger Sperry's laboratory at the California Institute of Technology. Most of the studies were collaborative efforts with Dr. Sperry and some with Colwyn Trevarthen.

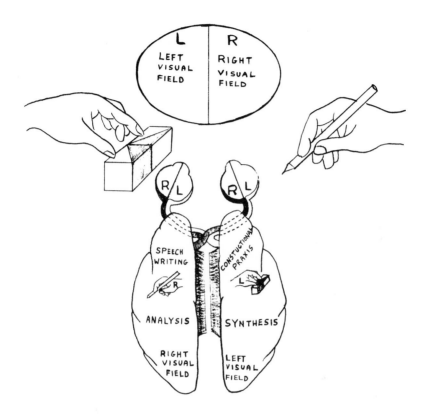

FIGURE 1. Schematic of the human brain after section of the corpus callosum.

which carry information to the right hemisphere, and vice-versa for stimuli in the right visual field. Any stimulus presented in the right visual field or to the right hand for tactual identification accordingly reaches the language hemisphere, and commissurotomy patients can then verbally name the stimulus. If a stimulus is presented in the left visual field or to the left hand, however, the information is conveyed to the mute right hemisphere, and the patients are unable to say what they saw or felt.

Repeated examination of nine commissurotomy patients has consistently confirmed this presence of strong lateralization and dominance in the left hemisphere of right handers for language and calculation (Sperry and Gazzaniga, 1966). Following surgery, the patient's behavior seems to be governed almost entirely from the more dominant leading left hemisphere, and in the great majority of tests conducted to date involving calculation, linguistic activities, or symbolic reasoning the left hemisphere is found to be far superior.

In contrast to the obvious superiority and dominance of the left hemisphere for the above functions, the corresponding specializations of the right, relatively mute, "minor" hemisphere have been much less easy to demonstrate. When we want to know what is going on in the left hemisphere, we have merely to ask the commissurotomy patient; but in the case of the mute minor hemisphere we are obliged to depend on special tests that utilize nonverbal forms of motor expression. There is a reluctance in some quarters to credit the mute illiterate minor hemisphere even with being conscious; the suggestion is that it is carried along in a trance-like automatic state with consciousness remaining unified and centered postoperatively in the dominant hemisphere. However, the evidence from a large number and variety of nonverbal tests strongly supports the idea that the minor hemisphere is indeed a conscious system in its own right—perceiving, feeling, thinking, and remembering at a characteristically human level.

More than this, it has been shown that the minor hemisphere is distinctly superior to the leading hemisphere in these patients in the performance of certain types of tasks—as, for example, in copying geometric figures, in drawing spatial representations, and in the assembling of Kohs blocks in block design tests (Bogen and Gazzaniga, 1965). The interpretation of these earlier observations remained uncertain in that it could not be determined whether the differential hemispheric capacities observed resided in the executive expressive mechanisms, as was suggested in certain aspects of the evidence, or whether they involved also more central perceptual and cognitive processing mechanisms.

In order to separate motor skills from perceptual processing, a test was devised which only required a simple motor read-out, namely pointing, but which required a rather complex understanding and manipulation of the spatial relationships (Levy-Agresti and Sperry, 1968). For this test a set of thirteen wooden blocks with three similar blocks in each set was constructed, each block differing from the other two within a set either in shape or in the relationship of textual surfaces. The patient felt one of the three blocks within a set with either his left or right hand, the hand hidden from view, thereby projecting the stimulus information to the right or left hemisphere, respectively. A card was then presented in free view to the patient on which was drawn two-dimensional representations of the three blocks in "opened-up" form. Figure 2 shows examples of two of the thirteen sets. The patient was required to point to the drawing which represented the block he was holding. It was thus necessary for the subject to mentally fold the drawings in order to select the correct match, and although both hemispheres could see the choice card, only one hemisphere knew which block was being felt.

A total of 156 trials was given per hand, each of the 13 sets being presented 12 times. Repeated presentations were possible

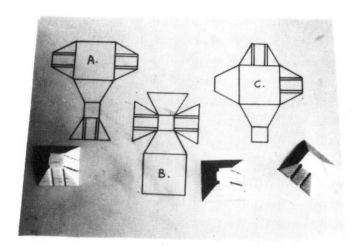

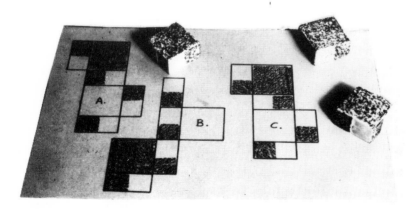

FIGURE 2.   Two examples of items from a cross-modal spatial relations test.

because the patients at no time ever saw the blocks, nor were they told whether a choice was correct or not. We saw no evidence of learning over the trials. The accuracy scores for the two hands on the twelfth replication were approximately the same as on the first.

A total of six patients was tested. Of these, two, both having severe right hemisphere damage, failed to even grasp the concept of matching a two-dimensional drawing to a three-dimensional object, even after 45 minutes of careful instruction. It should be pointed out that a normal seven-year-old child understood the test immediately and performed with a high degree of accuracy. Of the other four patients, one, also having right hemisphere damage, performed at chance level with both hands. The results of the other three patients were all similar in that their left hands were superior to their right at a high level of significance. Two of these were at chance level with their right hands, but above chance with their left. The other was above chance with both hands, but vastly superior with the left.

In addition to the quantitative superiority of the minor hemisphere, a qualitative difference in performance in several respects was noted. When the left hand was feeling a block, responses were quite rapid. On the other hand, when the right hand was feeling a block, the patients often took as much as 45 seconds to respond. In addition, when the right hand was feeling a block, there was a tendency for the patient to verbalize, saying such things as "A square, two rough sides, next to each other." It was difficult for us to inhibit such verbalizing. It was also noted that the sets which were relatively easy and difficult for one hand were not necessarily those which were easy or difficult for the other hand. Since each set had been presented to each hand a total of twelve times, it was possible to derive a score for each of the thirteen sets and to run a correlation between these scores for each hand. Interestingly, we found that the correlation between left-hand scores of different subjects was higher than between the hands of the subject who had above chance scores with both hands. In other words, the right hemispheres of different people found the rank ordering of difficulty for the thirteen sets to be more similar than the two hemispheres of the same individual. Examination of those sets which were relatively easy or difficult for one or the other hemisphere, led to the conclusion that the hemispheres processed the information in entirely different ways. Figure 3 shows the two sets which showed the largest disparity of difficulty for the two hemispheres. Set 7 yields itself to fairly simple analytic descriptions, but not easily discriminable visualizations. Set 2 contains figures which would be rather difficult to differentially describe, but which yield themselves to easily discriminable visualizations. It appeared that while the left hemisphere tried to

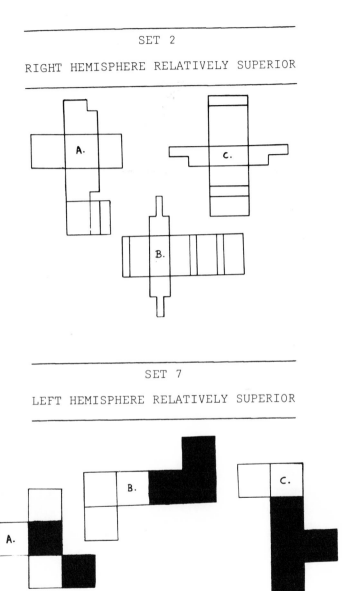

SET 2

RIGHT HEMISPHERE RELATIVELY SUPERIOR

SET 7

LEFT HEMISPHERE RELATIVELY SUPERIOR

FIGURE 3.   The two items showing the largest disparity of difficulty for the two hemispheres.

solve the problem by means of verbal-symbolic analysis, the right hemisphere utilized simple visualization. The major hemisphere seemed to be unable to break away from the verbal-analytic mode.

In work initiated in April of last year in collaboration with Roger Sperry and Colwyn Trevarthen (Levy, Trevarthen, and Sperry, 1972), we changed the basic paradigm of our tests. Instead of testing the two hemispheres separately and comparing their respective performances, we devised a method by which either hemisphere is free to respond. Kinsbourne and Trevarthen in unpublished work had found that when a stimulus such as a square is presented in midline of the visual field of commissurotomy patients, each hemisphere, rather than perceiving the half-square which is actually projected to the hemisphere, perceives a complete square. There is, in other words, an hallucinated completion of the stimulus by each half-brain. Utilizing this completion phenomenon, we presented chimeric stimuli in midline consisting of the left half of one stimulus joined to the right half of another stimulus. Figure 4 is an example of a chimeric face. Such a stimulus, presented in midline to a split-brain patient, is not perceived as a chimera. In fact, such questions as, "Did you notice anything odd about what you saw?" invariably produced a puzzled expression and the statement that he saw nothing strange about it. A stimulus such as this is perceived as one face by one hemisphere and as another face by the other hemisphere. We presented different kinds of chimeric stimuli for 150 m second in a tachistoscope while the patient fixated on a midline point. His task was then to point to the one of a set of non-chimeric whole stimuli presented in free vision which represented what he saw. Our results consistently showed that for faces, bisymmetric non-sense figures, line drawings of objects, and patterns of crosses and squares, the patients overwhelmingly pointed to the choice stimulus which represented what he had seen in the left half of the visual field; that is, what had been seen by the right, minor hemisphere. Only when we had the patients name what they had seen, rather than pointing to a matching stimulus, did they respond to the right-field stimulus, and for faces and non-sense shapes verbal responses were barely above chance level. In other words, for those stimuli whose names had been only recently learned, the major hemisphere had great difficulty in matching a stimulus with its name. It should be pointed out that the patients typically took 10 to 15 minutes to learn the names of the three faces and the names of the three non-sense shapes prior to the verbal naming tests. They learned the three names in about a minute, but seemed to be unable to connect these names with faces. Eventually they only learned the names by saying such things as "Dick has glasses, Paul has a moustache, and Bob has nothing"; in other words by noting analytic details. It was clear that the major hemisphere suffered a Gestalt perceptive deficit. In another type of test when we presented drawings of object chimeras and told the patient to point to a choice which was similar to what he saw, we found very interesting results. In this test "similarity" could mean either conceptual similarity like an eating utensil and a cake or structural similarity like a cake on a cake plate and a hat with a brim. The stimuli used for chimeric presentation and the choice stimuli are illustrated in Figure 5. When the right hemisphere responded, that is, when he pointed to a choice which was similar only to the left-field stimulus, he pointed to a choice which was structurally similar. When the left hemisphere responded, he pointed to a conceptually similar object. On any given trial one of the three possible responses would have been doubly correct, that is, similar in some respect to the stimuli in both half fields. For chimeric pairs 1-2, 3-1, and 2-3, the doubly correct response would have matched the right-field stimulus on a conceptual level and the left-field stimulus on a

FIGURE 4.   A chimeric face.

visual level. For chimeric pairs 2-1, 3-2, and 1-3, the doubly correct response would have been the reverse, that is, left-field conceptual and right-field visual matches. In fact, 24 out of the 27 doubly correct matches were visual matches with the left field and conceptual matches with the right field.

The results of this test showed the visual versus conceptual modes of information processing very clearly and showed unambiguously the association between hemispheres and modes of matching. However, these findings were with a single patient, and other patients we have tested have only shown conceptual matches, all of these with right-field stimuli. Even when we instructed these people to "Point to the thing which looks similar to what you see," they continued to make conceptual right-field matches. This may be simply due to a strong left-hemisphere dominance for matching behavior of objects related by similarity. In other words, it may be that the left hemisphere is more able to deal with similarity as a concept than is the right hemisphere, while the right hemisphere is dominant for identities.

Another test also involved object chimeras, in this case chimeras of a rose, an eye, and a bee. The patients were shown as choice stimuli drawings of toes, a pie, and a key. Neither the names of the stimulus nor the choice objects were ever spoken aloud and the subjects were told to "Point to the picture whose name rhymes with the name of what you see." Although we had already found that with simple recognition the patients recognized preferentially the left-field pictures, in this "rhyming objects" test, they invariably pointed to the object which rhymed with the right-field picture. This test clearly demonstrated that where recall of auditory images associated with visual stimuli was required, it was the left, major hemisphere which performed the task.

In contrast to the earlier tests, these chimeric tests not only show a difference in quantitative and qualitative capacity of the two hemispheres, they also show that the hemisphere which is superior for a function assumes control of the motor read-out. In these tests we had the patients point with the right and left hands and found no difference in the responses. When a particular test involved capacities for which the minor hemisphere was best equipped, the right hand as well as the left pointed to the left-field stimulus. The results from these tests represent the first demonstration of minor hemisphere dominance for motor control in commissurotomy patients. Of theoretical interest here is, "How does a disconnected hemisphere know it is superior?" Possibly an inferior hemisphere, confronted with a difficult task, simply makes no attempt to act. Or possibly the midbrain attention mechanism, receiving inputs from the two hemispheres, selects the superior

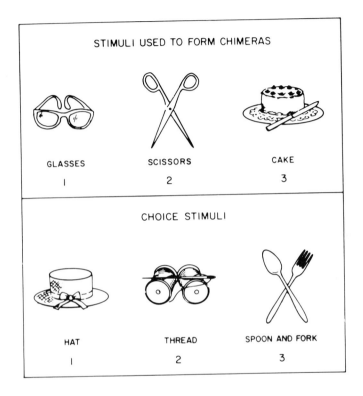

FIGURE 5.   Structure-function matching test.

hemisphere for the given task, and selectively "turns on" that hemisphere. We have no clear evidence either way as of yet.

In summary, our studies show that the disconnected minor hemisphere is the superior and dominant brain for perceptual recognition of faces, non-sense figures, pictures of objects, patterns, in detecting structural similarity, but not conceptual similarity, and in performing perceptual transformation, as well as for the motor read-out which communicates the results of its processing. The major hemisphere is the dominant and superior brain for verbal communication, for recognizing conceptual similarities, for recalling the auditory images of visual objects, and for the noting of analytic details.

Considering these various results, I was led to the idea that perhaps there is an intrinsic antagonism between the modes of information processing of the two hemispheres. Perhaps a hemisphere which is capable of expressing itself in language does not merely have the capability of symbolic-analytic reasoning, but is, in fact, constrained to use such reasoning. Such a hemisphere thinks in terms of symbolic and not visual relationships. It tends to look at stimuli analytically, as if it intended to describe what it sees. If true, this idea provides a basis for understanding why in man, but in no other animal, there is such a profound functional differentiation of the two half brains. Once the ancestral hominids acquired the capacity for language, there would be an obvious adaptive advantage for that capacity to be confined to a single hemisphere, leaving the other free to carry on the perceptive Gestalt functions. Had both sides of the brain possessed language, there would have been a loss in visualization abilities. The old adage, "He can't see the forest for the trees" possibly expresses most succinctly the proposed antagonism. Perhaps a language-competent hemisphere is simply incapable of seeing forests.

Fortunately for research purposes, whatever evolutionary mechanism led to the development of functionally specialized hemispheres, was not perfect. Some ten or so percent of people, in fact, do not have the highly specalized hemispheres we see in the majority. Left-handers tend to have some language competency in both halves of the brain. A cerebral lesion in either hemisphere more often leads to aphasia in sinistrals, but the aphasia is transient. If, after recovering from aphasia, a left-hander subsequently suffers an injury in the previously undamaged hemisphere, he again becomes aphasic, but this time permanently. It would appear that the left-handed individual, though typically having one hemisphere which is language dominant, nevertheless possesses a brain, either half of which is competent to handle language functions. The existence of this group of people provides the possibility for testing the hypothesis of a language-perception antagonism. If the hypothesis is true, people with bilateral language mechanisms ought to show a perceptual deficiency.

Two studies in the literature by Silverman and colleagues (1966) and William E. James and colleagues (1967) suggest that such a deficit might exist. Silverman found sinistrals to be poorer than dextrals in adjusting a luminescent rod to the absolute vertical in a darkened room where the rod was surrounded by a tilted luminescent frame. They were also poorer at locating a point of tactile stimulation on the dominant hand and at identifying pictures of various body parts as being left or right. Silverman suggested that perhaps left-handers are similar to children in having poorly differentiated hemispheres and that since children are relatively poor on the rod and frame test, a measure of so-called "field dependency," lack of differentiation is correlated with being field dependent. However, the fact that the same correlation exists in both left-handers and children is not an explanation of the basis of the correlation. In addition, neither tactile localization nor identification of left-right body parts is a test of field dependency. Silverman also pointed out that a greater percentage of brain-damaged people are left-handed than in the population in general. This results from the fact that damage to certain regions of the left hemisphere causes a genotypic dextral to become a phenotypic sinistral. Since about 65 percent of genotypic sinistrals have language-dominant left hemispheres, the same injury in these people does not cause a switch in handedness, but presumably merely switches the motor control of the left hand from ipsilateral to contralateral pathways. The net result is that if brain damage occurs with equal frequency in the left and right hemispheres, the proportion of observed left-handers among the brain damaged will be increased. On the basis of these findings, Silverman and colleagues suggested that among left-handers in the general population, a larger proportion suffer from unrecognized brain damage than right-handers. It is not, of course, a logical necessity that because there is a large proportion of sinistrals among the brain damaged that there is a large proportion of brain damage among sinistrals. In addition, the amount of unrecognized brain damage sufficient to produce a switch in handedness in the population at large must be rather small, and, in view of the large genetic component in handedness, it seems unlikely that the brain-damage hypothesis accounts for the deficiencies in sinistral performance. There is also the fact that the postulated genotypic dextrals who are phenotypic sinistrals as a result of unrecognized brain damage would presumably have suffered a lesion in their left, language hemispheres since it is only a lesion in this half of the brain which would cause a right-hander to become left-handed. This being the case, one would expect a verbal, rather than a perceptual deficiency, in a group of brain-damaged left-handers.

James and colleagues found sinistrals to be poorer than dextrals on a test of spatial orientation and on tests of perceptual closure. These investigators suggested that the deficit was due to mechanical disadvantages of using the left hand in making test responses. This appears to be an insufficient explanation since the differences between left- and right-handers were large and the motor responses were rather simple.

I compared the verbal and performance IQs of left- and right-handed graduate science students at Caltech on the Wechsler Adult Intelligence Scale (Levy, 1969b) since verbal and performance scores on this test have been found to measure what are normally left and right hemisphere abilities respectively. Although Verbal IQs were not significantly different for the two groups, 138 for the dextrals and 142 for the sinistrals, Performance IQs were 130 for the former and 117 for the latter, a highly significant difference. The Verbal IQ-Performance IQ discrepancy was 8 points for the right-handers and 25 points for the left-handers, a difference in the two groups which was significant at less than the .0002 level. Robert Nebes (1971 a, b) confirmed this performance deficit with undergraduates at Duke and with high-school students. Nebes devised a test consisting of tactually perceivable segments of circles in which a subject feels the segment and then selects the whole circle from which the segment comes. He found that the left hands of commissurotomy patients were superior to their right hands in performing this task, and that left-handers were poorer than right-handers, again illustrating that tasks for which the right hemisphere is superior are performed poorly by left-handers.

It seems that the best explanation for the foregoing results is that bilateral language competency does, in fact, produce a specific interference with Gestalt perception by compelling the two hemispheres to process information in a piecemeal, analytic, and symbolic manner which is incompatible with Gestalt apprehension.

It is interesting that the perceptual deficit seen in left-handers is also present in females in general. Using the Porteus Maze Test, Porteus (1965) found, in testing dozens of cultures all over the world ranging from that of Australian aborigines to that of French school children, girls were significantly inferior to boys. MacFarlane Smith (1967), in his book, *Spatial Ability,* has also pointed out that females show a specific spatial disability. It might be that female brains are similar to those of left-handers in having less hemispheric specialization than male right-hander's brains. A recent report by Culver and associates (1970) that right- as well as left-handed females show a greater primary amplitude of evoked responses in the right hemisphere than in the left lends support to the idea that the cerebral mechanisms responsible for perceptual deficiencies are similar in women and sinistrals since an earlier study by Eason and others (1967) found the same effect only in men who were left-handed. It is hard to reject the notion that a spatial-perceptive deficit in women is a sex-linked, genetically determined incapacity, an incapacity which possibly results from hemispheres less well laterally specialized than those of males. That the sex chromosomes do participate in determining spatial ability is given strong support by the finding that girls with Turner's syndrome, an XO condition, have a profound defect in spatial perception (Alexander, Ehrhard, and Money, 1966).

In work still in progress at the University of Colorado (Levy and Mandell, 1971) the nature of cerebral organization in left-handers is becoming clearer. Some sinistrals write with the left hand inverted, that is, with the pencil held above the line of writing, pointing toward the body. Others write in the normal fashion, although with the left hand. It has been generally thought that inverted writing in left-handers is a deliberate, conscious attempt to compensate for left-to-right writing with the left hand. However, I have observed a person who is an ambilateral who writes with his right hand, with the right hand inverted. His right hand has been the writing hand since first grade and he never showed any tendency to write with his left hand. Such inversion in a right-handed writer cannot, of course, represent any kind of compensation since there is nothing to compensate. This individual showed the typical perceptual deficiency of left-handers: his Verbal IQ was 34 points higher than his Performance IQ. Based on this single intriguing observation, I theorized that the manner of writing in sinistrals might reflect which hemisphere was language dominant. Most left-handers, though having less well-differentiated hemispheres than right-handers, nevertheless usually have one hemisphere which is the more dominant for language. As mentioned, approximately 65 percent of left-handers have a language-dominant left hemisphere and 35 percent a language-dominant right hemisphere (Goodglass and Quadfasal, 1954). I guessed that inverted writing in sinistrals reflected ipsilateral motor control from a left language-dominant hemisphere, whereas normal writing reflected the typical contralateral control seen in right-handers, but in this case, from a language-dominant right hemisphere. If this is true, we should see differences in inverted and normal writers on certain types of tasks.

Doreen Kimura (1970) has reported that when very brief exposures of verbal stimuli are presented in the right visual field of right-handers, recognition accuracy is superior to left-field exposures, but that when a single dot is exposed in any one of twenty-four positions in the left field, dot location accuracy is superior to that for dots exposed in the right field. In other words, a verbal task yields superior performance when projected to the language-dominant left hemisphere of dextrals, whereas a perceptual task yields better performance when projected to the mute right hemisphere. She interprets these findings in the light of the known differences in hemispheric function.

Mandel and Levy (1971) have repeated Kimura's experiments with left-handers who write either in an inverted or normal fashion to compare their responses. This investigation is still in progress, but so far we have found that inverted writers, like right-handers, are superior for dot location when dots are projected in the left visual field, suggesting that the right hemisphere of these people is superior for perceptual tasks. Normal writers among the left-handers show no field differences in accuracy of dot recognition, suggesting a lack of hemispheric specialization with respect to perceptual ability. On the other hand, inverted writers show no field differences in accuracy of word identification, indicating a lack of hemispheric specialization for verbal functions, whereas normal writers show superior word recognition for words projected in the left visual field, indicating a language-dominant right hemisphere. In other words, inverted writers appear to have specialized hemispheres with respect to perception but not with respect to verbal functions, whereas normal writers appear to have no hemispheric specialization with respect to perceptual functions, but a right hemisphere dominance for language functions.

A word of caution should be mentioned with respect to the above interpretations. Some investigators have found a left-field superiority for words even in right-handers. This result typically occurs when words are projected simultaneously in both fields and is interpreted to be a function of left-to-right reading habits. However, it is possible for this contaminating factor to operate even when words are projected only in a single field on a given trial, depending on various aspects of the experimental design which remain unclear. This contamination can either be minimal, in which case a right-field superiority appears, or it can be maximal, in which case a left-field superiority appears, or it can occur to such a degree that it counterbalances the hemispheric specialization effect, in which case no field differences for words are observed. It may be that in the present experiment the lack of field differences in word recognition accuracy for the inverted writers reflects just such a contamination which counteracts a right-field superiority which otherwise would have appeared. So far, we have only tested thirteen right-handers, but from this limited data it does appear that such a contamination is present in that there is no significant field difference in word recognition for this group either. If this holds up, then since it is known that dextrals in the vast majority have language-dominant left hemispheres, the lack of field differences for words in the inverted left-handers should be interpreted to reflect a language-dominant left hemisphere since their performance is identical to that of right-handers. It would then follow that inverted or normal writing in sinistrals reflects left and right hemispheric dominance for language, respectively, and right-hemisphere dominance for perception in the former group and no perceptual dominance in the latter group. Both these groups of left-handers had a perceptual deficit as measured by the Performance Scale of the WAIS, and there was no difference in the degree of the deficit in the two groups. This finding suggests that irrespective of which hemisphere is language dominant in sinistrals, the dominance is incomplete, producing an interference with perception. It thus appears that degree of lateralization and direction of lateralization are to some extent, independent.

In sum, the data from both studies of commissurotomy patients and from studies comparing left- and right-handers confirm the idea that there are two modes of information processing, each specific to a single hemisphere in the typical right-hander, that these modes are antagonistic when occurring together in the same hemisphere as in left-handers, and that the evolutionary reason for lateral specialization is explained by this antagonism.

ACKNOWLEDGMENT: The research reported here was supported by the Frank P. Hixon Fund of the California Institute of Technology and by the United States Public Health Service Grant No. MH-03372 to Roger W. Sperry at the California Institute of Technology and No. MH-46980-01 to the author at the University of Colorado.

Questions and Answers

QUESTION: You said that the way a left-hander holds a pencil has to do with the brain. Well, I'm a school teacher, and they tell us to place papers on a child's desk so that the lower left hand corner points toward him. If you do this with a left-handed child he writes with his hand bent around, but some left-handed children turn the paper around so that the lower right hand corner points toward them and these children write normally. So how can you say it has to do with the brain?

ANSWER: I wouldn't deny the possibility of environmental influence on the way a child holds a pencil. However, I would have to see the results of a carefully controlled experiment demonstrating such an environmental effect before I concluded that it existed.

In the cases you mentioned it is quite conceivable that it is those children with a right dominant hemisphere who turn the paper placed on their desks and then proceed to write normally, while those with left hemisphere dominance leave the paper as placed by the teacher, and then write with the hand inverted. Cause and effect are totally confounded in your observations. If the nature of left-handed writing were merely a function of how a piece of paper is placed on a desk, then one is left with the problem of explaining the perceptual differences observed by Mandel and myself in the two types of sinistrals, as well as the problem of explaining why the right-handed writer I observed writes with his right-hand inverted.

QUESTION: Understanding is that prehistoric skulls have been seen in which the two sides of the cranial case were unequal in size, the left being bigger. Yet these skulls came from creatures who weren't even *Homo sapiens.* You said that functional differentiation began when man developed language so how do you explain the observations of the skulls?

ANSWER: I'm not familiar with these observations, but they are unrelated to my hypothesis. I never maintained that language was uniquely an ability of *Homo sapiens,* but rather that it was a feature of Man. By Man I mean to refer to the Hominid family. In fact, it is my guess that language developed long before even *Homo erectus* appeared. Dart deduced from examination of the skulls of baboons killed with an antelope bone by Australopithecus over a million years ago, that somewhat over 90% of these primitive ape-men were right-handed, and it seems not unlikely that they also possessed the beginnings of a true language.

QUESTION: I've read that man developed a preferred hand because he was a tool-user, and that it was tool-using which caused his brain to develop the way it did. What do you think of this idea?

ANSWER: First, the presence of unilaterality, that is, of a preferred hand, does not induce hemispheric laterilization. Mice and rats have a preferred paw, but functionally symmetric hemispheres. Secondly, it is apparent that a preferred hand does not depend on being a tool user since mice are not tool users. Thirdly, even if an animal has a preferred hand and in addition is both a tool user and tool maker, as is the case with chimpanzees, it does not imply hemispheric specialization. Finally, pawedness, or handedness is not genetic in animals, but is in man, which would suggest that handedness in man is a secondary effect of cerebral dominance. If there's a selective advantage to having a preferred hand rather than ambilaterality, this advantage is conferred on all mammals who possess, not a genetically determined superior paw, but a genetically determined ability to *develop either* paw as the superior one. This latter endowment would have been totally sufficient for man as well if all he had needed were a major and minor hand. In fact, such a genetically endowed plasticity would have had greater selective advantage than the more rigid one we humans possess since damage to a genetically superior hand results in great difficulties in training a genetically inferior one. Since there would be no advantage and possibly a disadvantage in having a preferred hand under genetic control, then the most reasonable conclusion, in my view, is that genetic control of handedness is merely a concomitant of genetic control of cerebral dominance. Therefore, it is putting the cart before the horse to suppose that cerebral dominance could result from a preferred hand, or that a preferred hand depends on tool using.

Literature Cited

Alexander, D., A. A. Ehrhard, and J. Money. 1966. Defective figure drawing, geometric and human in Turner's syndrome. J. Nerv. Ment. Dis., *142:*161-167.

Arnett, M. 1964. A model of the inheritance of handedness and cerebral dominance. Nature, *204:*59-60.

Arnett, M. 1967. The binomial distribution of right, mixed, and left handedness. Quart. J. Exp. Psychol., *19:327-333.*

Bogen, J. E., and P. J. Vogel. 1962. Cerebral commissurotomy in man. Preliminary case report. Bull. Los Angeles Neurol. Soc., *27:*169-172.

Bogen, J. E., and P. J. Vogel. 1963. Treatment of generalized seizures by cerebral commissurotomy. Surg. Forum, *14:* 431-433.

Bogen, J. E., E. D. Fisher, and P. J. Vogel. 1965. Cerebral commissurotomy: A second case report. J. Amer. Med. Assoc., *194:* 1328-1329.

Bogen, J. E., and M. S. Gazzaniga. 1965. Cerebral commissurotomy in man: Minor hemisphere dominance for certain visuospatial functions. J. Neurosurg., *23:* 394-399.

Broca, P. 1960. Remarks on the seat of the faculty for articulate language followed by an observation of aphemia. In *Some Papers on the Cerebral Cortex,* G. von Bonin, tr., Springfield, Ill.: C. C. Thomas.

Collins, R. L. 1968. On the inheritance of handedness. I. Laterality in inbred mice. J. Hered., *59:* 9-12.

Collins, R. L. 1969. On the inheritance of handedness. II. Selection for sinistrality in mice. J. Hered., *60:* 117-119.

Critchley, M. 1962. Speech and speech-loss in relation to the duality of the brain. In *Interhemispheric Relations and Cerebral Dominance,* V. B. Mountcastle, ed., pp. 208-213. Baltimore: Johns Hopkins Press.

Culver, C. M., J. C. Tanley, and R. G. Eason. 1970. Evoked cortical potentials: relation to hand dominance and eye dominance. Percep. Mot. Skills, _____: 407-414.

Eason, R. G., P. Groves, C. T. White, and D. Oden. 1967. Evoked cortical potentials: relation to visual field and handedness. Science, *156:* 1643-1648.

Ettlinger, G. 1964. Lateral preference in the monkey. Nature, *204:* 606.

Finch, G. 1941. Chimpanzees' handedness. Science, *94:* 117-118.

Goodglass, H., and F. A. Quadfasal. 1954. Language laterality in left-handed aphasics. Brain, *77:* 521-548.

Hécaen, H. 1962. Clinical symtomatology in right and left hemisphere lesions. In *Interhemispheric Relations and Cerebral Dominance,* V. B. Mountcastle, ed., pages 215-243. Baltimore: John Hopkins Press.

Hécaen, H., and R. Angelerques. 1962. Agnosia for faces (prosopagnosia). Arch. Neurol. (Chicago), *7:* 92-100.

James, W. E., R. B. Mefferd, Jr., and B. Wieland. 1967. Percep. Mot. Skills, *25:* 209-212.

Kimura, Doreen. 1970. Presentation at the APA symposium, "Asymmetrical functioning of the human brain." Miami, Florida, September 4, 1970.

Levy, Jerre. 1969a. *Information Processing and Higher Psychological Functions in the Disconnected Hemispheres of Human Commissurotomy Patients.* Doctoral dissertation, California Institute of Technology.

Levy, Jerre. 1969b. Possible basis for the evolution of lateral specialization of the human brain. Nature, *224:* 614-615.

Levy, Jerre, and J. Mandel. 1971. Lateral field differences for verbal and nonverbal material in sinistrals. Study in progress.

Levy, Jerry, C. Trevarthen, and R. W. Sperry. 1972. Perception of bilateral chimeric figures following hemispheric deconnection. Brain, 95: 61-78.

Levy, Jerre, and C. Trevarthen. 1972. Hemispheric specialization tested by simultaneous rivalry for mental associations. Paper in progress.

Levy, Jerre and T. Nagylaki. 1972. A model for the genetics of handedness. Genetics, in press.

Levy-Agresti, Jerre, and R. W. Sperry. 1968. Differential perceptual capacities in major and minor hemispheres. Proc. N.A.S., 61: 1151 (abstract).

Nebes, R. 1971a. Superiority of the minor hemisphere in commissurotomized man for the perception of part-whole relations. Cortex, 7: 333-349.

Nebes, R. 1971b. Handedness and the perception of part-whole relationship. Cortex, 7: 350-356.

Paterson, A., and O. L. Zangwill. 1944. Disorders of visual space perception associated with lesions of the right cerebral hemisphere. Brain, 67: 331-358.

Petersen, G. M. 1934. Mechanisms of Handedness in the Rat. Comp. Psycol. Monograph No. 46.

Piercy, M., and V. O. G. Smyth. 1962. Right hemisphere dominance for certain nonverbal intellectual skills. Brain, 85: 775-790.

Porteus, S. D. 1965. Porteus Maze Test, Fifty Year's Application. Palo Alto, Calif.: Pacific Books.

Silverman, A. J., G. Adevai, and W. E. McGough. 1966. Some relationships between handedness and perception. J. Psychosom. Res., 10: 151-158.

Smith, I. M. 1967. Spatial Ability. San Diego, Calif.: Robert R. Knapp.

Sperry, R. W., and M. S. Gazzaniga. 1966. Language following surgical disconnection of the hemispheres. In Brain Mechanisms Underlying Speech and Language, F. L. Darley, ed. New York: Grune and Stratton.

Trankell, A. 1955. Aspects of genetics in psychology. Amer. J. Hum. Gen., 7: 264-276.

Van Wagenen, W. P., and R. Y. Herren. 1940. Surgical division of commissural pathways in the corpus callosum; relation to spread of an epilectic attack. Arch. Neurol. Psychiat., 44: 740-759.

Thirty-second Annual Biology Colloquium

Theme: The Biology of Behavior

Dates: April 13-14, 1971

Place: Oregon State University, Corvallis

Special Committee for the 1971 Biology Colloquium: John A. Kiger, chairman; Ronald H. Alvarado, Victor J. Brookes, Lyle R. Brown, Robert E. Larson, Ralph S. Quantrano, Carol A. Saslow.

Standing Committee for the Biology Colloquium: Ernst J. Dornfeld, Paul R. Elliker, Henry P. Hansen, Hugh F. Jeffrey, J. Kenneth Munford, Robert W. Newburgh, Paul O. Ritcher, J. Ralph Shay.

Colloquium Speakers, 1971:

> B. W. Agranoff, Department of Biological Chemistry and Mental Health Research Institute, University of Michigan, leader
>
> John Fentress, Biology Department, University of Oregon
>
> George L. Gerstein, Department of Physiology, University of Pennsylvania School of Medicine
>
> William D. Kaplan, Division of Biology, City of Hope Medical Center
>
> Jerre Levy, Department of Psychology, University of Colorado
>
> Arnold J. Mandell, Department of Psychiatry, University of California at San Diego

Sponsorship:

> Environmental Health Sciences Center, Oregon State University
>
> Graduate School, Oregon State University
>
> Phi Kappa Phi
>
> Research Council, Oregon State University
>
> School of Agriculture, Oregon State University
>
> School of Home Economics, Oregon State University
>
> School of Humanities and Social Sciences, Oregon State University
>
> School of Science, Oregon State University
>
> Sigma Xi